Vegetation of the Soviet polar deserts

Studies in Polar Research
This series of publications reflects the growth of research activity in and about the polar regions, and provides a means of disseminating the results. Coverage is international and interdisciplinary: the books will be relatively short (about 200 pages), but fully illustrated. Most will be surveys of the present state of knowledge in a given subject rather than research reports, conference proceedings or collected papers. The scope of the series is wide and will include studies in all the biological, physical and social sciences.

Vegetation of the Soviet polar deserts

V.D. ALEKSANDROVA
TRANSLATED BY D. LÖVE

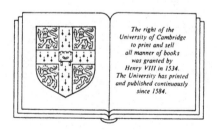

The right of the
University of Cambridge
to print and sell
all manner of books
was granted by
Henry VIII in 1534.
The University has printed
and published continuously
since 1584.

CAMBRIDGE UNIVERSITY PRESS

CAMBRIDGE

NEW YORK   NEW ROCHELLE   MELBOURNE   SYDNEY

CAMBRIDGE UNIVERSITY PRESS
Cambridge, New York, Melbourne, Madrid, Cape Town, Singapore, São Paulo, Delhi

Cambridge University Press
The Edinburgh Building, Cambridge CB2 8RU, UK

Published in the United States of America by Cambridge University Press, New York

www.cambridge.org
Information on this title: www.cambridge.org/9780521329989

Originally published in Russian as *Rastitelnost poliarnykh pustyn SSSR* by Nauka, Leningrad, 1983 and © Nauka 1983

First published in English by Cambridge University Press 1988 as *Vegetation of the Soviet polar deserts*

English edition © Cambridge University Press 1988

This digitally printed version 2008

A catalogue record for this publication is available from the British Library

Library of Congress Cataloguing in Publication data

Aleksandrova, V. D. (Vera Danilovna)
Vegetation of the Soviet polar deserts.
(Studies in polar research)
Translation of: Rastitelnost poliarnykh pustyn SSSR.
Bibliography
Includes index.
1. Desert flora – Soviet Union, Northern.   2. Desert flora – Arctic regions.
I. Title.   II. Series. QK321.A48513   1988   581.974′2   87-17886

ISBN 978-0-521-32998-9 hardback
ISBN 978-0-521-10005-2 paperback

Contents

Translator's foreword

When you step ashore for the first time in a high-arctic country, the immediate impression is one of bleakness and desolation. You wonder if any life can exist there. It takes close inspection and much energy to discover the scanty vegetation which actually manages to survive in a climate where life-giving heat is reduced to a minimum and the stature of the plants, hidden among the stones, often can be measured in centimetres only.

Vera Danilovna Aleksandrova is one of the outstanding Russian botanists who has devoted much time and energy to exploring the vegetation on the Soviet arctic islands in high latitudes. The present publication is a review of the vegetation of the arctic polar deserts within the USSR in general with the main emphasis on results obtained during Dr Aleksandrova's study of the island of Zemlya Aleksandra in the Zemlya Frantsa-Iosifa archipelago. This area was thoroughly examined and the author has analysed the correlation between the vegetation, the climate, including the microclimate, the substrate and the insolation. A large number of relevés revealed in detail the composition of the flora of both higher and lower plants in relation to the ecological conditions. This resulted in a composite diagram of the ecological position and amplitude of 12 different plant aggregations and 12 leading indicator species. The latter part of the book is devoted to a phenological study of the day-to-day development of certain vascular plants. It must have taken an unusual patience and persistence to measure the few millimetres of growth per day of these plants but the findings clearly convey an impression of the struggle for survival that goes on at those high latitudes. Because of the extreme sensitivity of the author for colour details, we learn also that in spite of the general bleakness there is much beauty in the landscape when you care to look for it.

For anybody concerned with ecological and phenological studies of vegetation under extreme conditions, this book is of great importance because of its wealth of detailed information.

The translation follows the original as closely as possible. The author herself has made a minor revision of the text (p. 205) and approved of the translation. The geographical names used are direct transliterations of the Russian epithets. The taxonomy is that of the author. In two cases only (pp. 53 and 191) I have added comments to help to explain more clearly the conditions observed. The term *plakor* (from the Greek word for 'flat, level' and denoting a zonal, mesic habitat) is also used here. It was first introduced in the English translation of another of Dr Aleksandrova's books: *The Arctic and the Antarctic: their division into geobotanical areas*, which was published in 1980 by Cambridge University Press.

I am indebted to my husband, Dr A. Löve, to Dr J. Major, University of California, Davis, CA, Dr W. A. Weber, University of Colorado, Boulder, CO, and especially to Dr S. W. Greene, University of Reading, Reading, England, for assistance with terminology and for general encouragement. I also want to thank my dear friend, Vera Aleksandrova, for her patience with my many questions and for our fruitful discussions relating to the translation. These have indeed been of great help. I am also grateful to Cambridge University Press for valuable advice and assistance throughout this work.

Doris Löve
San José, CA, USA

Preface to English edition

Vera Danilovna Aleksandrova, now retired from the Academy of the Sciences although still available for advice to students and fellow workers, has devoted almost her entire scientific career to botanical exploration of the Soviet arctic islands, especially Zemlya Frantsa Iosifa and Novaya Zemlya and also some of the East-arctic islands and the tundras of Yakutia. Her life now spans 8 decades and after her arduous explorations, often under severe and dangerous conditions, she resides with her family in Leningrad.

This book treats in considerable detail the vegetation of the Arctic polar deserts. Its focus is primarily on polar deserts of the Barents and Siberian provinces, no account being given of the vegetation of polar deserts in the third province, the Canadian, which stretches from the north of Canada to the north of Greenland. By concentrating almost exclusively on the work of Russian scientists, and integrating their results with much original data collected by the author herself, the book presents the first comprehensive account, in the English language, of the plants and vegetation of the Soviet arctic polar deserts, north of the Eurasian landmass. A valuable report of the author's observations of seasonal development of flowering plants in localities of Zemlya Frantsa Iosifa (Frans Josef Land) is also included.

When presenting her classification and account of the vegetation, the author discusses relevant concepts of vegetation description developed by Lavrenko, Norin, Senninkov, Sukachev and others and introduces some new terminology. Since many of the concepts and terminology used by Russian scientists to describe vegetation are not well known or understood outside of the Soviet Union, the reader may find it helpful to consult V. D. Aleksandrova's (1973) English language paper cited at the end of this preface.

For the present English edition an extra figure (the Map, pp. xii) has been inserted which shows the position of the more important of the localities mentioned in the text. These are as given in *The Times Atlas of the World* (comprehensive edition, 1983). The artwork used in the production of this book is the original used for the Russian edition.

It is a pleasure to thank Dr Löve for the care she has taken with the translation, Drs F. Daniëls (Utrecht) and D. W. H. Walton (Cambridge) for advice with the phytosociological terminology, and Dr T. E. Armstrong (Cambridge) for help with the transliteration of Russian names.

ALEKSANDROVA, V.D. 1980. Russian approaches to classification of vegetation. In Whittaker, R.H. (ed.). *Classification of plant communities.* The Hague, W. Junk B.V., Publishers, 167-200.

S. W. Greene, December 1986

Author's acknowledgements

When writing this book I was able to draw fully from the instruction and counsel of N. V. Matveyeva, B. N. Norin, T. G. Polozova, I. N. Safronova, O. I. Sumina, E. A. Khodachek and B. A. Yurtsev. To all those mentioned as well as to A. L. Abramova, N. S. Golubkova, A. L. Zhukova, M. Lamb, A. N. Oksner and R. N. Shlyakov, who determined the cryptogams in my collection from Zemlya Frantsa-Iosifa, I wish to express my deep gratitude.

Abstract

This book is an account of the vegetation of the far northern areas of the Soviet Arctic. It includes data collected during research on high-arctic areas that have already been published as well as original material collected by the author during her work on Zemlya Frantsa-Iosifa. The peculiarities of the soils and the microclimate are described and a list of the plants is provided. The structure of the plant aggregations is analysed, and a classification of them is worked out. The high-arctic territory is divided into areas on the basis of the characteristics of the plants within the geographical boundaries of the USSR. The peculiarities of the biomorphology and the characteristics of the seasonal development of the plants in the arctic polar deserts are described on the basis of investigations made by the author. The book presents arguments for distinguishing the area of the arctic polar deserts as a special geobotanical region.

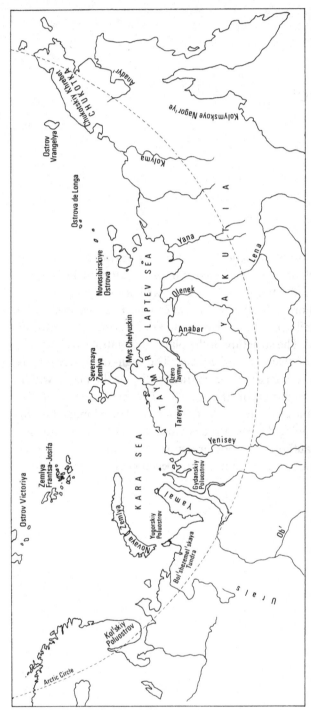

Map showing the location of the major place names mentioned in the text.

1

Introduction

The polar deserts of the USSR are situated within the area of the botanical region of circumpolar arctic polar deserts.

The epithet 'desert' ('cold desert') was first used by Passarge (1920) as a designation of the territories extending across the highest latitudes, where major portions of land as well as of sea are covered by ice all the year round, and where, in general, the yearly radiation balance amounts to only 10 kcal/cm^2 and where the mean temperature of the warmest month does not exceed 2°C. The application of the term 'desert' was used because the vegetation in the parts of this region not covered by ice is very poor: it is absent from most of these areas and where it does occur it exists only in the form of scattered, more or less large fragments in among bare parts of the mostly stony substrate or as patches, separated from each other. The vegetation consists mainly of lichens (among which crustose types predominate) and mosses, together with rare and very small specimens of flowering plants; a more or less continuous cover is met with only under more favourable conditions, where a major role is also played by lichens. The 'desert'-like aspect of this territory is caused by the extreme lack of heat (whereas in tropical and subtropical deserts the major factor is the extreme lack of water).

The essential differences between 'desert' areas of the high latitudes of the Arctic and the tundra territory bordering them to the south, where the vegetation cover is considerably richer, constitute the basis for the distinction of the former into a high-latitude zone different from the tundra. Berg (1928, 1936) called it the 'frozen' zone, Gorodkov (1935) the 'nival' zone and Leskov (1947) the 'high-arctic nival area'. Gorodkov (1935, p. 127) suggested that the type of vegetation typical of the 'nival zone' of the Arctic should be called 'arctic deserts'.

The fact that (in the Southern Hemisphere) in the Antarctic there is vegetation which is similar to that of the arctic deserts with respect to both life forms and the structure of its associations is the main reason (according to e.g. Aleksandrova, 1963, 1977*a*; Korotkevich, 1967, 1972) for adopting a broader term, i.e. 'polar deserts'.

Within the region of the arctic polar deserts two wide belts can be distinguished, i.e., a northern and a southern, as well as three provinces, i.e., the Barents province, the Siberian province and the Canadian province (Aleksandrova, 1977*a*). Within the northern belt the vegetation is very impoverished, the flora of flowering plants is poor, and its representatives belong to arctic–alpine types; species with a hyparctic type of distribution are completely absent. Within the southern belt the number of flowering plants is somewhat greater (see Table 6): in the most favourable local conditions, plant associations can be encountered which are typical of those of the northern belt of the arctic tundra; among these there are occasionally various representatives of hyparctic species.

The polar deserts of the USSR are situated within the areas of the Barents and the Siberian provinces (Fig. 1.). They occur on the archipelago of Zemlya Frantsa-Iosifa, Ostrov Victoriya, Ostrova Vize, Ushakov and Uyedineniya, the northern tip of Novaya Zemlya, the archipelago of Severnaya Zemlya, Mys Chelyuskin, and the Ostrova De Longa. Arctic polar deserts outside the USSR occur on Nord-austlandet and Kong Karls Land in Spitsbergen, bordering on the Barents province; the northernmost group of the islands in the Queen Elizabeth archipelago and a narrow coastal belt of Ellesmere Island in the Canadian north; and in Greenland a coastal belt north of the area of distribution of *Carex stans*.

Tedrow (1972) and Bliss (1977) place the limit of the polar deserts in North America further south, including the area where the territory displays alpine polar deserts as well. However, the polar deserts distributed on heights within those areas appear to be the result of vertical belt formation within a part of a territory belonging to the subzone of arctic tundra (see p. 207).

The complex characteristics of the natural conditions within the major part of the region of arctic polar deserts, including the characteristics of the vegetation, have been discussed by Korotkevich (1972) in his well-known monograph on polar deserts. Although recently published books mentioned in the Bibliography concerning the flora and the vegetation of the Soviet polar deserts (e.g. Tolmachev and Shukhtina, 1974; Safronova, 1976, 1979, 1981*a*, 1981*b*, 1983; Aleksandrova, 1977*b*, 1981; Blagodatskikh, Zhukova and Matveyeva, 1979) furnish new information, there appears to be a need for the publication of more details on aspects of the

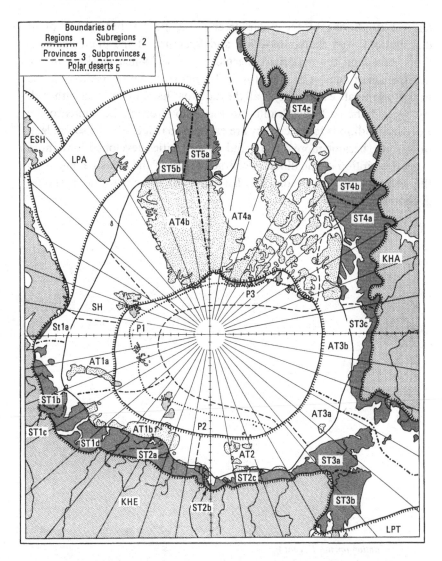

Fig. 1. The limits of Arctic geobotanical areas. The key indicates the borders
and boundaries between: *1*, geobotanical regions; *2*, subregions; *3*, provinces; *4*,
subprovinces; *5*, the northern and southern belts of the polar deserts. Regions:
LPA, the Northatlantic grass- and heath-land; *ESH*, the European broadleaved
forests; *KHE*, the Eurasiatic coniferous forests (taiga); *LPT*, the Northpacific
grass- and heath-land; *KHA*, the North American coniferous forests (taiga).
Subregions of the tundra area: *ST*, subarctic tundra; *AT*, arctic tundra. Provinces of the subarctic tundra area: *ST1*, East European–West Siberian subarctic tundras (subprovinces: *ST1a*, Kola; *ST1b*, East European; *ST1c*, Urals-Pai-
Khoy; *ST1d*, Yamal–Gyda–Zapadnyy Taymyr); *ST2*, the West Siberian subarctic tundras (subprovinces: *ST2a*, Khatanga–Olenek; *ST2b*, Kharaulakh; *ST2c*,
Yana–Indigirka); *ST3*, the Chukotka–Alaska subarctic tundras (subprovinces:

natural conditions of the polar deserts, the pecularities of the plant distributions, and the composition and structure of the plant aggregations, based on the material collected by myself when working in the archipelago of Zemlya Frantsa-Iosifa. Therefore it seems expedient for me to try to generalize and analyse all the data that have so far been gathered. As a result the pathways leading to the adaptation of the plants to the extreme conditions to which they are exposed, particularly the severe lack of warmth, became more distinct and opportunities developed for a more exact description of the different characteristics of the areas belonging to the arctic polar deserts and the distinction of these areas as a special geobotanical region.

Caption for fig. 1 (cont.)

ST3a, Chukotka; *ST3b*, Anadyr'–Penzhina; *ST3c*, Alaskan); *ST4*, Canadian subarctic tundras (subprovinces: *ST4a*, Keewatin; *ST4b*, Hudson; *ST4c*, Labrador); *ST5*, Greenlandic subarctic tundras (subprovinces: *ST5a*, West Greenland; *ST5b*, East Greenland). Provinces of the Arctic tundra area: *AT1*, Novaya Zemlya–West Siberian – Central Siberian arctic tundras (subprovinces: *AT1a*, Novaya Zemlya – Vaygach; *AT1b*, Yamal– Gyda–Taymyr–Anadyr'; *AT2*, East Siberian arctic tundras; *AT3*, Ostrov Vrangel'–West American arctic tundras (subprovinces: *AT3a*, Vrangelian; *AT3b*, West America); *AT4*, the north-east Canadian–North-west Greenlandic arctic tundras (subprovinces: *AT4a*, North-east Canada–North Greenland; *At4b*, North-east Greenland). *SH:* the Spitsbergen autonomic geobotanical district. Provinces within the region of the polar deserts: *P1*, the Barents; *P2*, the Siberian; *P3*, the Canadian.

2

General characteristics of the landscape

Orography and glaciology

The islands on which the vegetation of the arctic polar deserts is distributed are the most elevated parts of a shallow continental shelf (which is less than 100–500 m below sea level), jutting up above the ocean surface by tens to hundreds of metres. The glaciated peaks on the northern island of Novaya Zemlya reach up to 1000 m a.s.l. (here and in the following, the elevation is always given as metres above mean sea level).

The archipelago of Zemlya Frantsa-Iosifa, Ostrov Victoriya and the islands in the northern part of the Kara Sea are parts of the submerged Barents plateau (Fig. 2), the base of which consists of precambrian layers reworked during subsequent periods of tectonic activity. This plateau was

Fig. 2. The Barents (*B*) and the Hyperborean (*H*) plateaux in the Arctic Ocean (according to the *Physical-geographical atlas of the World*, 1964, p. 5) *1*, the coastline of the present continent; *2*, the limit of the continental shelf; *3*, the boundaries of the plateaux.

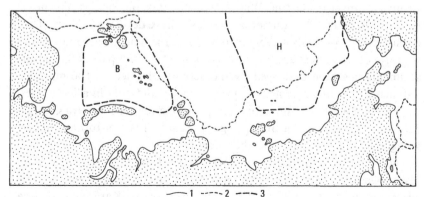

separated from the old Pan-arctic plateau (occupying almost all of the Arctic) by the younger Caledonian and Hercynian orogeneses but was also eroded and submerged and could thus reappear during the present epoch of the Neogen period. The rest of the Pan-arctic plateau appears in the form of the Hyperborean plateau (Fig. 2). Ostrova De Longa rise above the ocean surface in its southern part.

Ostrov Victoriya, the westernmost of the Soviet arctic islands, is the tip of one of the peaks on the Barents plateau. The total surface of this island amounts to 5.3 km²; 98% of it is covered by an ice sheet which reaches an altitude of 100 m a.s.l. Only a narrow belt in the northern part of the island is free of ice. It forms Mys Knipovich, which consists of an accumulation of marine sediments and boulders (Govorukha, 1970*b*).

The archipelago of Zemlya Frantsa-Iosifa consists of 186 islands, more than 85% of the surface area of which is covered by ice; parts of the ice edge are afloat on the sea. The land area which is free of ice is largely made up of capes and nunataks. Only five islands have ice-free areas of considerable size: Zemlya Georga, Ostrov Greem-Bell, Zemlya Aleksandra, Zemlya Vil'cheka and Ostrov Kheysa. Their ice-free areas amount to 490, 464, 150, 148 and 84 km², respectively.

From a geological point of view, Zemlya Frantsa-Iosifa represents a major portion within the area of the Barents plateau that is free of neo-tectonics. The archipelago is situated on a rigid base, built mainly of mesozoic sedimentary layers, with a lower cretaceous basaltic cover and early quaternary fragments broken into a variety of blocks, so that at present the islands represent an entire system of horsts and grabens; because of this the greatest differences in height between its elements above and below sea level amount to 1000 m. The highest point within Zemlya Frantsa-Iosifa (the Wullersdorf nunatak on Zemlya Vil'cheka) reaches 670 m a.s.l. The quaternary deposits have developed slowly. They consist mainly of sandy-pebbly (rarely clayey) sediments, accumulated in the form of marine terraces but also as coastal banks. A part of the marine sediments has been destroyed by the activity of glaciers. The morainic sediments are not very thick, and occur rarely and only in a few places, almost all of which are found on Zemlya Aleksandra and Zemlya Georga. Like Ostrov Victoria, the archipelago of Zemlya Frantsa-Iosifa is undergoing glacio-isostatic uplift at the present.

The islands in the northern part of the Kara Sea (Ostrova Vize, Ushakova and Uyedineniya) are also peaks on the Barents plateau (or, according to a different terminology, the Barents-Kara plateau (see Tkachenko and Atlasov, 1970)). They are composed of mesozoic and quaternary layers on top of a folded precambrian base. Ostrov Ushakova is completely covered by ice; its highest peak reaches almost 350 m a.s.l. There are no glaciers on Ostrova Vize and Uyedineniya. The not very extensive lowlands of these two islands reach elevations of up to 25 m a.s.l. and have an erosion–accumulation type of relief; the mesozoic bedrock is represented by sands, and the quaternary elements by marine sandy-pebbly sediments and moraines in the form of sands with boulders. The south-western shores of these islands, especially that of Uyedineniya (Sisko, 1970), are characterized by thermo-abrasive disintegration.

The far northern part of Novaya Zemlya consists of the northernmost outposts of the Novaya Zemlya mountain range, which were built up within the Uralo-Hercynian folded system. Within the area belonging to the arctic polar deserts (Fig. 1), the major part of the territory is occupied by the northern part of an ice sheet the altitude of which reaches 1000 m a.s.l. here. Its tongues of discharging glaciers fill the bays with icebergs. The shores of the bays are characterized by ice barriers. The land by the shore which is free of ice occurs only in narrow strips; in other parts it is reduced to nothing. Only in the vicinity of Mys Zhelaniya in the north-eastern part of the island does it widen out: here it is a ridged and hilly flatland with boulder-strewn morainic sediments cut by rivers and brooks which originate from the ice sheet (Zubkov, 1934; Samoylovich, 1937).

Severnaya Zemlya consists of an archipelago of more than 30 islands, four of which are large: Ostrov Oktyabr'skoy Revolyutsii, Ostrov Bol'she-vik, Ostrov Komsomolets and Ostrov Pioner. The major part consists of smaller islands, including: Ostrova Shmidt, Malyy Taymyr, Staroka-domskogo, Krupskoy and Naydenysh. Severnaya Zemlya represents a part of the Taymyr–Severnaya Zemlya folded area with complex layers, which include in part precambrian, Caledonian and Hercynian structures and in part mesozoic ones composed of metamorphosed sediments and magmatic bedrock. The main role is played by paleozoic sandstones and, among the igneous rocks, by granites. The old folded paleozoic layer underwent in its time a denudation, leading to a peneplanar surface. The plateau thus formed was fragmented into blocks by the action of cenozoic dislocation, followed by uplift and subsidence, leading to the present configuration of the archipelago.

The remnants of the old peneplane appear as parts of a plateau, the altitude of which ranges from 200 to 800 m a.s.l., which constitutes the main surfaces of Ostrova Bol'shevik and Oktyabr'skoy Revolyutsii (Semenov, 1970). Secondary geomorphological elements are sections of gently rolling plains, with elevations ranging from 120 to 200 m a.s.l.; these are best developed in the central part of Ostrov Oktyabr'skoy Revolyutsii. A third geomorphological element is a low-lying flatland some 100–200 m a.s.l., which is met with on all the islands (Semenov, 1966a, 1966b, 1970). On this plain the basal bedrock is covered by marine and glacial quaternary sediments, mainly represented by stony loams with boulders.

Ice in the form of ice sheets and ice domes, the highest point of which reaches 950 m a.s.l. (Govorukha, 1976), covers about 50% of the surface of the archipelago (Semenov, 1981a). Floating glaciers give rise to icebergs. In the north-eastern part of Ostrov Oktyabr'skoy Revolyutsii, in the fjord of Matusevich, shelf ice occurs, formed by glaciers originating from fans discharged by the Karpinskiy and Rusanov sheets. At the present time Severnaya Zemlya is experiencing uplift, which is connected not only with glacio-isostatic effects but also with tectonic causes; because of the latter subsidence of various sections (e.g., the eastern shores of Ostrov Komsomolets) is occurring simultaneously with the general uplift.

Just like Severnaya Zemlya the area of Mys Chelyuskin belongs to the Taymyr–Severnaya Zemlya folded area. As stated by Matveyeva and Chernov (1976) the relief there is formed by low ridges of bedrock outcrops (clayey and calcareous siltstone predominate, more rarely there are broken down quartz veins or dolomite outcrops) alternating with concave sections where the bedrock is covered by quaternary marine sediments in the form of medium or rather heavy loams. There are no glaciers.

As already mentioned, Ostrova De Longa are situated on the southern part of the Hyperborean plateau (see Fig. 2) near its edge. The Hyperborean plateau, which has been strongly broken up by tectonic actions, is at the present time uplifted above or submerged below the surface of the ocean in the form of a mosaic of major blocks; some of these blocks are distinguished by folded formations of synclinal type or by wide zones of broken or disturbed areas (Tkachenko & Atlasov, 1970). On Ostrov Bennett sediments from the Cambrian period have been preserved in the form of argylites with strata of limestone and siltstone. At the end of the early Cretaceous a thick layer of basal bedrock was formed within the area of Ostrova De Longa. On Ostrov Bennetta there is a series of layers of basaltic porphyrites and on Ostrov Genrietty and Ostrov Zhannetty there are tufogenic strata (Sisko, 1970). All of these three northern islands, Bennetta, Genrietty and Zhannetty, are covered by glaciers. On the more southerly island, Ostrov Zhokova, there is 123-m tall peak, apparently build up of basalt, but on the small plain of Ostrov Vil'kitskogo there is no glacier. The glaciers on Ostrov Bennetta are indicated by three ice domes with elevations of up to 426 m a.s.l., which cover about half the surface of the island (Kartushin, 1963a). On Ostrov Genrietty there is a small ice sheet, which reaches more than 300 m a.s.l. in elevation. There is also a small ice dome (total surface 0.2 km^2) on Ostrov Zhannetty, the top of which was found to reach an altitude of 350 m a.s.l. and the ice cover of which measures 80 m in thickness.

This brief review demonstrates that glaciation is a prominent feature of the islands within the Barents province of the arctic polar deserts: ice covers the major part (about 90%) of the land surface and nourishment for the glaciers is assured by the moisture within the atlantic air masses. It has been suggested that at the time of the Pleistocene glaciation there was a single Barents Sea ice sheet covering all the northern part of the Barents Sea shelf, including Spitsbergen and Zemlya Frantsa-Iosifa. The presence on Mys Knipovicha on Ostrov Victoriya of pebbles and boulders, the structure of which reveals that they originated from Spitsbergen and Zemlya Frantsa-Iosifa, speaks in favour of this hypothesis (Govorukha, 1970*b*). There is also the opinion (Dibner, 1970) that during the early Würm period there was a Taymyr–Kara ice sheet extending over the islands of Severnaya Zemlya and covering parts of the Kara Sea all the way to the eastern border of the Ste. Anne trench, where an 'interglacial corridor' between the Barents Sea ice sheet and the Taymyr–Kara sheet existed. In the Siberian province of the polar deserts, where there is less assurance of moisture than in the Barents province, glaciers at present cover about one half of the land surface.

A major factor in the history of the formation of the present landscape was the extended length of the Pleistocene period, when the islands were united with the continent and there was only one land mass, i.e., during the maximum emergence of the shelf. The period of shelf emergence was in general contemporary with the Ice Age, during which a colossal amount of moisture was concentrated in the ice cover. However, the glacio-isostatic uplift of the continent during interglacial periods was also of importance for the emergence of the shelf. Regressions of the sea alternated with transgressions; at the time of the latter the continental sand bars were again flooded. During the maximum regression of the sea its surface fell below the present level by 100–200 m, and maybe by up to 500 m. This resulted in the existence for a long period of time of an enormous expanse of dry land on the continental shelf, at the edge of which the presently submerged but geomorphologically well-defined belt of the continental slope was formed and here and there cut by canyons. Its boundary can be seen on the map of the Arctic Ocean published by Dibner *et al.* (1965). The marine transgression at the time of which the surface level of the sea exceeded the present level by 100–200 m, as well as the regression following it, played major roles in the formation of the landscape's relief (i.e., the formation of abraded and accumulated terraces, coastal banks, etc.) and of the quaternary sediments. The transgression destroyed any existing vegetation, just as the glaciation had. At the time of the last glaciation, those territories which are at present occupied by arctic polar deserts were completely covered by an ice sheet; there was no refugia and the colonization by vegetation started only when the glaciers declined and areas of periglacial land which were not covered by ice appeared.

Climate

The climatic characteristics of the areas which belong to the arctic polar deserts are determined first and foremost by their high-latitude position: both the length of the polar night and the low position of the sun during the long polar day influence the climate. The climate is also affected by the presence of a large quantity of ice with high albedo and low temperature both in the ocean and on the land with the result that the radiation balance is very low, being less than 10 kcal/(cm^2 year) (see the *Physico-geographical atlas of the world*, 1964). The temperature during the vegetative period is also constantly low: the mean July temperature does not exceed 2°C.

The circulation of the atmosphere and the existence of sea currents are also of great importance for the development of the climate. In particular, differences in the climates of the Barents and Siberian provinces of the arctic polar deserts are brought about by this. Intense cyclonal activity and the penetration far towards the north of a branch of the Gulf Stream cause a distinct expression of a marine type of weather within the Barents province. This is characterized by frequent fogs, very dense cloud cover (the frequency of occurrence of cloudy skies amounts to 90% over Zemlya Frantsa-Iosifa), frequent changes in the weather pattern, the presence of 'mild' winters and a heavy snow cover in comparison to that of the Siberian province. In areas of the latter province an anticyclonal circulation predominates during the winter in connection with the East Siberian anticyclone and the anticyclone activity over the eastern portion of the arctic basin. Cyclones occur during the summer, but when they do, they abate rapidly. As a result the climate within the Siberian province of the polar deserts does not show such extreme effects of marine characteristics as that of the Barents province, taking on instead traits of a moderate continentality. While the amount of precipitation on Zemlya Frantsa-Iosifa is about 300 mm each year and the thickness of the snow cover on surfaces close to the sea averages 40–60 cm (Govorukha, 1970a), the amount of precipitation within the Siberian province amounts to no more than 100–230 mm and the thickness of the snow cover does not exceed 30–40 cm (Semenov, 1970).

As an illustration of the distinct continentality of the climate within the Siberian province of the polar deserts a comparison can be made between the characteristics of the climate in the area of Bukhta Tikhaya on Zemlya Frantsa-Iosifa and that of Mys Chelyuskin (Prik, 1970). During winter the type of weather with hard frost and temperatures from −33°C to −42°C that is characteristic of Mys Chelyuskin does not occur at Bukhta Tikhaya, although the latter is situated much farther north. On Mys Chelyuskin the absolute minimum temperature is 10°C lower than that at Bukhta Tikhaya and during the summer there are more days with cloudless weather. Thus not only the value of the absolute minimum but also that of the absolute maximum temperatures as well as the difference between winter and summer weather are much more significant on Mys Chelyuskin (Figs. 3 and 4).

Areas with a cryo-arid climate are not found within the limits of the Soviet arctic polar deserts. They occur in the Canadian province of the polar deserts, where the total yearly precipitation, according to Porsild (1955), amounts to only about 75 mm, and according to Bliss (1975) to 50–150 mm. Based on data from King Christian Island, the total precipitation during the summer amounts to 35–40 mm (Bell and Bliss, 1978) as a consequence of which plant aggregations are best developed in those habitats where they receive moisture from melting snow. We should also remember that the southern belt of the antarctic polar deserts, situated on the Antarctic Continent, differ by more extreme cryo-aridity (with the exception of the northwestern part of the Antarctic Peninsula where, just as on the adjacent islands, the climate is cryo-humid, like the climate on Zemlya Frantsa-Iosifa (see Aleksandrova, 1977a)).

A review of the regional differences of the climates within the arctic polar deserts of the northern hemisphere is illustrated in a diagram published by Korotkevich (1972, p. 134) showing the division of the Arctic into climatic areas.

When considering the climate of this region it is necessary to remember also that the winds are extremely strong in all the areas of the arctic polar deserts. Cold wet winds have a great effect on the formation of the climate when they are blowing in from an ocean where there is always floating ice. Similarly it is necessary to stress the local importance of foehn winds, which arise due to the presence of high mountains and deep valleys within fjords. The occurrence of foehn winds such as are recorded on Severnaya Zemlya (Semenov, 1970; Korotkevich, 1972) is characteristic within the area of the Soviet arctic polar deserts. They also occur on Ostrov

Bennetta. When a foehn wind develops, the temperature rises sharply while the relative humidity of the air falls rapidly. The rise in temperature can be quite significant: up to 15°C, and sometimes 20°C or more, within a few hours. In the more southerly areas, but still within the area of the arctic tundra, at sites where locally repeated foehn winds occur there are extrazonal representatives in the vegetation cover such as thickets of hyparctic dwarf shrubs in the valleys of the internal parts of Ostrov Vrangelya or thickets of *Salix lanata* in the valley below the eastern slopes

Fig. 3. Structure of the climate and weather in the area of Mys Chelyuskin (according to Prik, 1970). Frost-free period: *1*, cloudless, moderately and very humid days; *2*, cloudy days without precipitation; *3*, cloudy nights without precipitation; *4*, cloudy, no precipitation; *5*, rain. Weather with temperatures hovering around 0°C (alternating frost and thaw): *6*, cloudless days; *7*, clear days. Freezing weather: *8*, light frost (0°C to −2°C) without wind; *9*, moderate frost (−3°C to −12°C) without wind; *10*, same as *9*, with wind; *11*, considerable frost (−13°C to −22°C) without wind; *12*, same as *11*, with wind; *13*, strong frost (−23°C to −32°C) without wind; *14*, same as *13*, with wind; *15*, severe frost (−33°C to −42°C) without wind; *16*, same as *15*, with wind; *17*, extreme frost (below −43°C) with wind. *a*, monthly amount of precipitation in mm; *b*, extreme values of maximum temperatures; *c*, graph illustrating the course of the temperature throughout the year; *d*, extreme values of minimum temperatures; *e*, duration (in days) of the period when daily temperatures are below 0°C and the mean dates of its beginning and end.

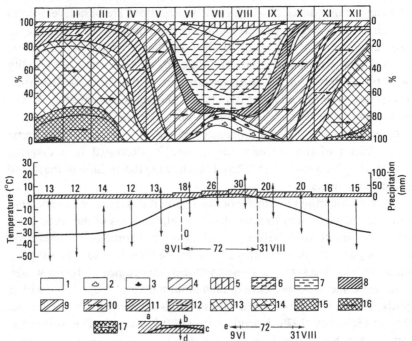

of the mountains of Byrranga on Taymyr peninsula (Dibner, 1961). On Severnaya Zemlya the occurrence of foehn winds is doubtless one of the factors promoting the development within the polar desert landscapes of 'oases' of arctic tundra vegetation containing species like *Salix polaris* (Korotkevich, 1958; Safronova, 1976).

Fig. 4. Structure of the climate and weather at Bukhta Tikhaya (according to Prik, 1970). Keys: the same as for Fig. 3.

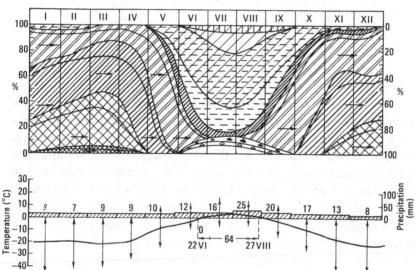

3

Conditions for the existence of vegetation

Substrates and soils

The substrate on which plants can become established within the arctic polar deserts consists of skeletal soils. This means most often stony elluvium or materials with boulders ranging from surfaces with very large boulders down to skeletal and gravelly-skeletal soils. Localities with loamy moraine deposits and marine sediments usually contain more or less mixture of gravel, stones and pebbles. The predominance of stony ground appears to be one of the *zonal* characteristics of the arctic polar desert landscape and is associated with the fact that erosion of the mountain rock base takes the form of fragmentation during the processes of physical weathering. Chemical and biochemical weathering processes are strongly suppressed due to the lack of heat, and because of this the weathering of the bedrock produces no clayey minerals. Substrates which contain clayey fractions are met with here and there, but they are not very thick and have usually been transported from somewhere else in the form of either moraine deposits or marine sediments rather than being developed on the spot.

Physical weathering processes are always intense because of the severe winters and the alternation between positive and negative temperatures during spring, summer and autumn. This factor leads also to the development of cryogenic types of surface formations. Freeze-sorting of substrates is especially widely distributed: the larger stones are frozen to the surface and if in the substrate there are any fine soils (small stones, gravel, fractions of sand or loam), these will be separated from the larger stones and formed into medallions surrounded by rings of stone or stones piled up in less regular configurations in the form of a stone net (Fig. 5). In those cases, which are generally rare, where the base of the ground – the origin of which is usually associated with more or less ancient clayey

marine sediments – consists of a clayey fraction, a substrate can form
which is broken into polygons by fissures. It is characteristic of the arctic
polar deserts that these are of small dimensions. On Zemlya Aleksandra
the diameter of the polygons within the zonal plakor aggregations
amounts to 30–50 cm, but in habitats with a late-disappearing snow cover
it can be only 10–30 cm. According to counts made by Matveyeva (1979)
about 400 polygons can be found within the better developed aggregations
of the polar deserts on Mys Chelyuskin, while in the arctic tundra there
are only 100–500 polygons per 100 m^2. Where the snow cover persists for a
long time, the number amounts to 1000 or more on Zemlya Aleksandra
according to counts that have been made. In Fig. 6, illustrating an
example from Zemlya Aleksandra, the polygons measure 10–25 cm in
diameter. On bare loam small patches of vegetation can be seen, formed
mainly of crustose lichens (*Pertusaria* spp., *Ochrolechia* spp., etc.) and
very small individuals of *Phippsia algida*.

The small dimensions of the polygons can apparently be explained by
the extremely slow seasonal movements of the permafrost due to the cold
weather and the fact that the fissures, which develop in the form of cracks
due to drying, start to form in a very thin layer of ground when the
permafrost has still not reached its maximum lower limit. Large tetragons,
such as appear as a result of frost cracking of ground and such as are
widely distributed in the continental areas of the arctic tundra zone, are
not found within the zone of the arctic polar deserts because of the lack of
an adequate thickness of fine soil sediments. Korotkevich (1972) has
described only traces of tetragonal polygons from Ostrov Oktyabr'skoy
Revolyutsii in the archipelago of Severnaya Zemlya.

On substrates with small-size polygons the shape of the polygons
approaches six-sided figures (Fig. 6). Narrow fissures, once they have
developed, are preserved for a long time. This determines the polygonal
nature of the surface on which the various elements of the vegetation
aggregations can become established where the most favourable habitats
for development exist and where, consequently, soil-forming processes
can start. Descriptions of soil profiles of arctic desert soils have been made
for sites with such substrates (e.g. Mikhaylov and Govorukha, 1962;
Korotkevich, 1972; Govorenkov, 1981). However, because of the frag-
mentary development of the vegetation cover, considerable areas are in
general not affected by soil formation.

Korotkevich (1972, p. 192) stated correctly: '... in connection with the
severity of the climatical conditions and especially the low temperatures,
the soil processes within the zone of the polar deserts are to a great extent
suppressed and extremely peculiar, which in turn leads to the formation of

individual types of soil not met with in the other geographical zones'. In contrast to those of the tundra, the soils of the arctic polar desert do not have any pronounced peat horizon; the latter occurs only in so-called 'foci' or 'pockets'. In addition, the soils of the Soviet arctic polar deserts are neutral or only weakly acidic and the soil complexes are almost completely saturated with moisture absorbed from the soil base (Mikhaylov & Govorukha, 1962; Mikhaylov, 1970; Korotkevich, 1972). Fulvic acids predominate in the humus. The lack of gleyization is always characteristic and a consequence of the low temperatures (in essence, this distinguishes the zonal polar desert soils from those of the tundra). Gleization can take place only in very favourable habitats and only in the southern belt of the polar deserts (Chugunova, 1979), where it can occur in the form of an extrazonal phenomenon. In the northern belt of the polar deserts rare cases of an ephemeral gleyization under a thin moss sward, saturated by moisture, have been recorded, although it was not

Fig. 5. Stone net with medallions of fine soil. On the island of Zemlya Aleksandra (Zemlya Frantsa-Iosifa) (photo.: the author).

visibly expressed in the soil profile but rather established by its reaction with potassium sulphocyanate (Govorukha, 1970*a*).

The composition of the soil microflora and soil fauna is specific. Actinomycetes, non-sporeforming bacteria and blue-green algae predominate in the soil microflora (Sushkina, 1960; Novichkova-Ivanova, 1963, 1972). Many species of soil fauna, typical of the tundra as well as of the arctic polar desert zone, occur, as demonstrated by the investigations made by Yu. I. Chernov (Matveyeva and Chernov, 1976) in the area of Mys Chelyuskin. The dominant groups are invertebrates (Nematodes, Enchytraeidae, Collembola, larvae of Chironmidae), but they are taxonomically very impoverished. A number of species are distinguished by their extreme, individual morphology. Very small forms predominate ('miniature life forms'). The trophic associations determined by the specificity of the plant aggregations have special characteristics.

Results of analyses of polar desert soils from Zemlya Frantsa-Iosifa, collected by myself and by L.S. Govorukha, have been published by

Fig. 6. Polar desert with polygons on the island of Zemlya Aleksandra (photo.: the author).

Mikhaylov and Govorukha (1962), of samples from Severnaya Zemlya by Mikhaylov (1960), Korotkevich (1972) and Govorenkov (1981) and of samples from the district of Mys Chelyuskin by Chugunova (1979).

Concerning provincial differences between the polar desert soils within the area of the Soviet Arctic, Mikhaylov and Govorukha (1962, p. 48) state: 'In contrast to the soils of Severnaya Zemlya and the Novosibirskiye Ostrova, those from Zemlya Frantsa-Iosifa are definitely more humid and form together with the soils of Spitsbergen an oceanic, atlantic faction of soils.' In his general treatise, Mikhaylov (1970) distinguishes the areas lying within the zone of the Soviet polar deserts into two soil provinces: the Atlantic and the Siberian. The polar desert soils within the territories of the Canadian Arctic Archipelago and Greenland have long since been distinguished as a special soil province. Because of the predominance there of basic soils and cryo-arid climatic conditions, the soils have an alkaline reaction and show a number of specific traits in respect to their chemistry and the morphology of the soil profiles (e.g. Tedrow, 1972). Soils of the Canadian province of the polar deserts are thus definitely different from the soils of the humid and sub-humid provinces of the Soviet arctic polar deserts. It is therefore impossible to extrapolate from the data on the soils of the North American polar deserts (Karavayeva and Targul'yan, 1978) to those of any other areas of the polar deserts.

Snow regime

The role of snow cover as an ecological factor within the arctic polar deserts is exceptionally important. According to data from Zemlya Frantsa-Iosifa (Govorukha, 1970*a*) the average thickness of snow cover in the atlantic areas of the polar deserts amounts at altitudes near sea level to around 0–40 cm; with increasing altitude the amount of snow precipitating increases. Less snow falls within the limits of the Siberian province of the arctic polar deserts according to data from Severnaya Zemlya (Semenov, 1970). At the end of the winter there the thickness of the snow does not exceed 30–40 cm.

Under the influence of the strong winds the snow is blown away from raised sites as well as from exposed parts of slopes and settles in depressions and low-lying parts of the slopes. Although not very extensive, its distribution over the topographical elements is specific, and areas exist that are similar to those in the tundra zone where, here and there, strong winds occur. However, the effect of snow cover on vegetation within the tundra and the polar deserts is essentially different. This is, in general, connected with a very much lower total amount of heat in the polar deserts, because of which the shortness of the vegetative period due to an only gradually disappearing snow cover can have catastrophic consequences for the vegetation. This does not occur in the tundra zone and the protective role of the snow, which is so distinctly evident in the tundra (especially in connection with the development of shrubs), drops to practically nothing in the polar deserts.

As demonstrated by my own investigations on the island of Zemlya Aleksandra (in the archipelago of Zemlya Frantsa-Iosifa), the most satisfactory development of the vegetation within the polar deserts occurs in localities with a thin snow cover, which are in part even entirely free of any snow during the winter, because, in spite of the severity of the winter conditions, the early disappearance of the snow from such localities is of primary importance. This means an extended vegetative period and a larger amount of total heat received by the plants. Thus rich aggregations consisting predominantly of lichens (see Table 10) can be encountered only where the snow is blown away completely or almost so or where it disappears at the earliest date possible.

The most favourable development of vegetation on this island occurs where the snow forms a layer 25 cm thick and disappears by 15–20 June. However, where the snow cover is more than 25 cm thick and disappears late or where the total amount of heat received by the plants is low, the vegetation is much poorer and becomes even more impoverished the thicker the snow cover is. Finally, in localities where the snow cover is more than 1 m thick and where during the year of my investigations the snow disappeared completely only after 25 July (this occurred on about 25% of the territory) there was no sign of any vegetation (Fig. 7).

These phenomena lead to a number of zonal characteristics of the arctic polar desert in its aspect as a special geobotanical region. In the tundra zone there is always – even where the snow disappears late – an adequately developed vegetation. The vegetation within the snowy tundra zone has been described by many authors. In the arctic tundra on the southern island of Novaya Zemlya there are depressions, occupied by mires, where there is a deep snow cover (Aleksandrova, 1956) and the ground is free from snow for only two months following the first thaw and where, in spite of this a rich mire vegetation has developed beautifully. One can even find plant aggregations in places below steep slopes where the snow lasts into the middle of August. Localities where there is no vegetation at all do not exist in the tundra zone. Vegetation-free localities begin to appear only in areas of increasing altitude in the mountains especially within the belt of the alpine polar deserts.

Data concerning the snow cover and the vegetation have been collected by myself on the island of Zemlya Aleksandra both in the form of visual observations along the route of investigation and in the form of a snow-measuring survey conducted along profiles and surfaces still covered by snow during the first days of June. A systematic mapping of the snow cover was made in terms of its disappearance and a description of the vegetation gradually emerging from beneath the snow.

An example of a profile which shows the changing depth of the snow cover from 2 June 1959 to 20 August 1959 and the emergence of vegetation as the snow disappeared is given in Fig. 8. Fourteen localities with different snow regimes were distinguished. The dotted line on the graph indicates the surface of the snow on 2 June. The characteristics of the localities are listed in Table 1.

The role of such snow drifts, which persist for several years and from which water flows during the summer, doubling the moisture available, is not important in the Barents province of the arctic polar deserts since there the summer is humid and the plants do not suffer from any shortage of water. In the Siberian province, however, as reported by Korotkevich

Fig. 7. Site without any sign of vegetation at a locality on Zemlya Frantsa-Iosifa with late disappearing snow (photo.: the author).

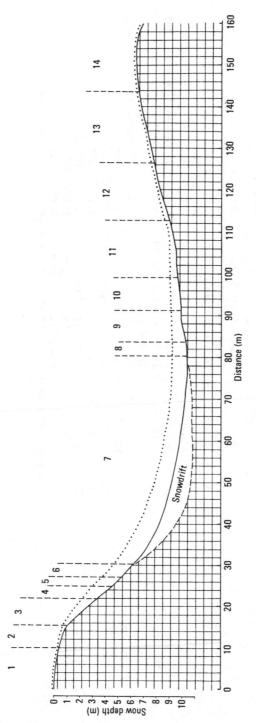

Fig. 8. A 160-m long profile on the island of Zemlya Aleksandra. *1–14* are the numbers of sites with different snow regimes (see Table 1); the characteristics of the sites are described in the text.

Table 1. *Characteristics of sites with different snow regimes as illustrated by the profile given in Fig. 8*

Site no.	Snow depth 2 June (cm)	Dates by which the snow had disappeared	Vegetation (see the numbers of the relevés in the tables indicated)
1	10	End of May to 5 June	32 (Table 12)
2	10–20	2 June to 12 June	33 (Table 8)
3	75–100	10 July to 16 July	55 (Table 15)
4	>100	24 July	56 (Table 18)
5	>100	25 July	57 (Table 20)
6	>100	7 Aug. to 9 Aug.	Bare ground with small rocks, no plants
7	>100	The snow had not melted by 20 Aug.	Permanent snow drift
8	>100	1 Aug. to 7 Aug.	Bare ground with small rocks, no plants
9	80–100	14 July to 24 July	66 (Table 16)
10	60–75	1 July to 10 July	61 (Table 16)
11	40–60	24 June	69 (Table 15)
12	20–45	15 June to 1 July	37, 38 (Table 15)
13	0–20	End of May to 10 June	36 (Table 12)
14	0–10	End of May to 5 June	34 (Table 12)

(1958), the plants often experience a lack of humidity and the moisture available from melting snow drifts is definitely of importance. It is even more so in the cryo-arid areas of the Canadian province of the polar deserts.

Microclimate

The temperature regime under which the plants of the polar desert develop depends on the fact that heat reaches the surface of the ground and is directly absorbed by the plants themselves in the form of direct and scattered solar radiation. The air masses can, as is well known, also absorb heat from the ground surface. Because of the fact that in areas near the pole localities on land not covered by ice occupy a very insignificant area, compared to the surface of the entire region where near the ground the air masses are affected by glaciers, floating ice and cold waters in the Arctic Ocean, the temperature of the air masses is extremely low and does not exceed a mean July temperature of 2°C at a height of 2 m within the region occupied by the polar deserts (as measured in the meteorological screens at the polar stations). However, rather than using data from the meteorological stations for the temperature regime of importance for the existence of plants growing near the ground, it is more satisfactory to refer not to air temperature data but to data on the temperature at the ground surfaces.

Graphs are furnished later (see Fig. 51) of the mean daily temperature in the air and at the ground on Zemlya Aleksandra. During the season of my investigation (in 1959) the mean daily temperature at the ground surface from the end of May to the end of August was always higher (by a magnitude of 0.1–0.4°C) than the mean daily air temperature. The importance of the daily maximum and minimum temperatures at the soil surface for the characteristics of the temperature regime relating to the existence of plants is highly significant: up to 15 July the minimum temperature at the soil surface was every day negative; from 27 July onwards, after a brief period when the minimum temperature was every day positive, it again began frequently to fall below 0°C. However, as repeated investigations (e.g. Aleksandrova, 1961) have demonstrated, the summer frosts do not severely damage the plants at high latitudes, although both their growth and the course of their phenological stages are hampered. Instead, the maximum temperatures, which the plants use for hastening their development, are of enormous importance. The excess of the maximum temperature over the daily mean, measured at the soil surface, amounted at some sites to 8°C. The maximum temperature at the soil surface became positive on Zemlya Aleksandra at the end of May, long before the mean temperature reached 0°C, and did not fall below freezing point until the end of August.

The circumstance that stones lying on the soil surface absorb more heat during the hours of maximum temperature than does the fine soil is very important for the temperature regime of the polar deserts. This can be explained by the fact that the fine soil, retaining more or less moisture, utilizes part of the heat for evaporation. It is true that stones lose heat faster during the part of the day when the temperature at the soil surface is lower on days with especially severe weather conditions but, as already stated, for the plants in the polar deserts the total amount of time at maximum temperature is most important. Stones also play a role in the appearance and extension of thawed patches: snow on sites with a thin snow cover begins to melt first around and above stones (Fig. 9) due to the effect of solar radiation penetrating the snow cover. In addition, favourably arranged stones can form a kind of 'stone hothouse' on the importance of which for the development of the plants more is said later.

Fig. 9. 'Greenhouses' and 'caves' of ice at the edge of patches thawing out on the island of Zemlya Aleksandra (Zemlya Frantsa-Iosifa) at the beginning of June (photo.: the author).

Table 2. *Temperature regime at the soil surface of thawed-out patches on Zemlya Aleksandra, 1959*

Date and time of day	Weather and condition of snow cover	Temperature (°C)			
		Air temp. at 2 m	At surface of stones, 10 × 20 cm	At surface of epigenic rhizoids of crustose lichens	Within 'ice greenhouses'
1 June, 4 p.m.	Snow depth 7–8 cm, melting around stones (Fig. 9), thickness of ice crust from 1–2 up to 5 cm	−4.5	Stones visible through thin layer of ice	—	—
2 June, 11.30 a.m.	—	−2.0	4	1.5–2.0	—
3 June, 11.30 a.m.	—	−1.5	2.5	0.5	0.5[a]
5 June, 4 p.m.	Strong northerly wind, blizzard	−3.5	0.5	—	0.5[b]
6 and 7 June	Blizzard, snow storm				
8 June, 3.35 p.m.	Snow storm abating, wind speed 6 m/s	−3.5	0.5–3.5	—	1.5 – 2
9 June, 11 a.m.	No precipitation, wind speed 3 m/s	−5.0	1–6.5	—	1–3.5
10 June	Same	−3.5	(2–3)–11.5	—	up to 5.5 1.5–7
14 June, 11.30 a.m.	Snow fell the day before but the snow was melting over the patch thawed out. Wind speed 2 m/s	1.0	9.0	—	6.5

Notes [a] In 'snow cave' above stones.
[b] In 'ice greenhouse' between stones.

My investigations on Zemlya Aleksandra (Table 2) have furnished examples concerning the role of stones and also of 'greenhouses' and 'caves' of ice (Fig. 9). Every day during my observations low clouds occurred (index 10). During the summer the difference in temperature between the stones and the soil surface was not so great but the stones were always warmer than the fine soil by at least 1–1.5°C.

The soil microclimate is important for the conditions of plant life (Fig. 10). Some data obtained by measuring soil temperatures at different depths on Zemlya Aleksandra are presented in Table 3; the observations were made at the beginning of June on some small thawed out patches in the vicinity of the polar station Nagurskaya. Later temperature measurements were made by means of thermometer probes at three points not far from each other at about 20 m elevation, where relevés 9, 11 and 17 were laid out.

In spite of the more satisfactorily developed vegetation, the permafrost below relevé no. 9 decreased only slowly, a fact which apparently can be explained as due to the heat-insulating effect of the moss–lichen sward which covered a higher percentage of the surface here than within other plant aggregations, although the temperature of the sward itself was in

Fig. 10. Daily course of air temperature and soil temperature in the polar desert of Mys Chelyuskin, 21–22 July 1974 (according to Chernov *et al.*, 1979). (*a*) Polygonal–stony loam with a 5–6% cover of lichens and mosses; (*b*) moss–lichen plant aggregations with a sward of vegetation, forming rows and cushions, with cover up to 20%; (*c*) lichen–moss plant aggregations with a sward of vegetation, distributed over a polygonal net, giving 60% cover; *I*, bare ground; *II*, lichen–moss sward. Temperature: *1*, of the air; *2*, of the soil surface according to a fixed thermometer; *3*, of the soil surface, minimum temperature; *4*, of the soil surface, maximum temperature; of the soil at depths of 5, 2 cm; 6, 5 cm; 7, 10 cm; 8, 20 cm.

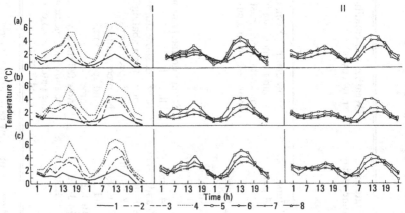

Table 3. *Temperature regime of the soil and depth of permafrost below thawed-out patches, Zemlya Aleksandra, 1959*

Date and time of day	Depth down to permafrost (cm)	Temperature (°C)		At 1–2 cm below the rhizoids of crustose lichens
		Air temperature at 2 m	At surface of rhizoids of crustose lichens	
2 June, 11.30 a.m.	1	−2	1.5, 2	0.5
3 June, 11.30 a.m.	2	−1.5	0.5	2–2.5
12 June, 3.15 p.m.	3	1	3, 4	—
13 June, 11 a.m.	6	0.5	4.5	3.5–4.5
14 June, 11.30 a.m.	10–13	1	6.5	5
16 June, 10 a.m.	16	0.5	4.5	4.5–5

Table 4. *Temperature regime of the soil, Zemlya Aleksandra, 1959*

No. of relevé; substrate; vegetation	Date	Depth down to permafrost (cm)[a]	Air temp. at 2 m (°C)	Soil temp. (°C) Below rhizoids of crustose lichens	Soil temp. (°C) Below moss–lichen turf	Soil temp. (°C) at a depth (cm) of 5	10	15	20	25	30
9; fine soil mixed with stones; the vegetation the most satisfactorily developed on the island; see Table 7	26 June	16	−0.2	3.5	—	3.5	1.5	—	—	—	—
	30 June	19	0.6	9.5	7.0	7.0	4.0	1.5	—	—	—
	4 July	25	−0.6	8.5	6.5	5.5	4.0	1.5	—	—	—
	9 July[b]	28	5.0	12.0	11.0	8.0	5.0	3.5	2.0	1.0	—
	14 July	32	1.8	8.0	7.5	6.0	5.0	4.0	2.5	1.5	—
	21 July[c]	35	—	—	—	2.5	2.5	1.5	1.0	0.5	—
	24 July	—	3.8	—	—	4.0	4.0	3.0	2.5	1.5	—
	30 July[d]	—	0.9	—	—	1.5	1.1	0.8	0.5	0.3	—
11; loam with a small amount of gravel and pebbles; zonal vegetation; see Table 9	30 June	24	0.6	8.0	—	6.5	4.0	2.0	—	—	—
	4 July	26	−0.6	8.5	7.0	5.5	4.0	2.0	0.5	—	—
	9 July[b]	28	5.0	12.5	10.5	8.0	5.5	4.0	3.0	1.5	—
	14 July	34	1.8	6.5	6.0	6.5	5.5	4.0	2.0	1.0	—
	21 July[c]	—	3.1	—	—	2.5	2.5	2.0	1.5	1.0	—
	24 July	—	3.8	—	—	4.5	4.0	3.5	3.0	1.5	—
	30 July[d]	—	0.9	—	—	1.5	1.2	1.0	0.5	1.0	—
	20 Aug.	39	2.5	6.0	4.0	5.0	4.0	2.5	2.0	1.5	1.0

17; loam with a
small amount of
small stones;
open aggregation
of type semi-
aggregation;
see Table 17

4 July	23	−0.6	—	5.0	3.5	2.5	0.5	—	—
9 July[b]	25	5.0	—	8.0	6.5	4.5	3.0	—	—
14 July	28	1.8	—	6.5	5.0	4.5	2.0	1.0	—
21 July[c]	30	3.1	—	2.5	2.5	2.0	1.5	0.5	—
24 July	—	3.8	—	6.0	5.0	4.0	3.0	0.5	—
30 July[d]	—	0.9	—	1.5	1.2	1.0	0.5	0	—
20 Aug.	38	2.5	—	5.0	4.0	2.5	2.0	1.0	0.5

Notes: [a] Mean of three measurements.
[b] Sunny weather; cloudy the rest of the day.
[c] On 20 June and during the night to 21 June, the weather was windy (wind speed up to 20 m/s).
[d] Temperature measured after cessation of a snowfall; lots of snow fell on 29 June and during the night to 30 June.

general higher. From the data presented in Table 4 it is apparent not only that the soil was very cold and that the seasonal lowering of the permafrost surface occurred slowly, but also that the soil temperature reacted quickly to changes in the weather (see the data for 9 June and 30 June!). Throughout August the temperature of the surface layer of the soil remained positive, but on 30 August, towards evening, puddles became covered by ice and during the night to 31 August the top layer of the soil froze firmly and was transformed into a solid, hard-frozen crust. On 31 August the puddles did not thaw out and ice formed along the shores of small lakes. On 1 September the top layer of the soil remained frozen.

Although the data presented in Table 4 are not fully representative, they all confirm that the temperature of the surface layer of the soil is higher than the temperature of the air during the daytime. This excess, such as recorded by myself, amounted to 1.5–9°C (and on an average to 4–6°C). Similar examples and results of measurements of soil temperatures were also obtained on Mys Chelyuskin (see Fig. 10). The plants strive to utilize the heat by pressing themselves to the ground surface (see the section on 'Seasonal development of the plants in the polar desert').

4

The vegetation

Data on the floristic composition of the plant aggregations of the arctic polar deserts in the literature are still very scanty. In his work on Severnaya Zemlya Korotkevich (1958) provided general characteristics concerning the vegetation of this archipelago and published lists of the specific composition of 14 relevés, limiting himself exclusively to an enumeration of the flowering plants and mentioning mosses and lichens only in connection with the total data on the estimated cover. Matveyeva (1979) has published lists of five relevés from the area of Mys Chelyuskin, including flowering plants, mosses, liverworts and the majority of the lichens (revealing not only the composition of the epilithic lichens but also their cover); in addition, the general characteristics of the vegetation of the area investigated and a short description of the communities met with there are included in her work. Short reports on the vegetation on Ostrov Bennet⁺⁹ are included in a paper by Kartushin (1963b), on the vegetation of Severnaya Zemlya (in addition to the above-mentioned work by Korotkevich) in papers by Safronova (1976, 1981b) and on the vegetation on Zemlya Frantsa-Iosifa in papers by Aleksandrova (1977b, 1979) and Safronova (1981a, 1983).

However, more detailed material, not yet published, has been collected by myself on the island of Zemlya Aleksandra (Zemlya Frantsa-Iosifa archipelago), where 70 geobotanical relevés were studied. Special lists of these are given below. A short review of existing data on the vegetation of the polar deserts in the Arctic area outside the USSR and in the Antarctic has been published (Aleksandrova, 1977a, pp. 133–4, 139–56; see also Aleksandrova 1980, pp. 152–68, 175–86).

Structural peculiarities of the plant aggregations within the polar deserts and the principles for their classification

The structure of the vegetation of the polar deserts on Zemlya Frantsa-Iosifa (Aleksandrova, 1977b) and on Mys Chelyuskin (Matveyeva, 1979) was investigated. The closed vegetation in these areas is limited in extent and the open type predominates; vegetation is completely absent from a significant portion of these territories.

One of the types of continuous vegetation occurs in the form of a moss–lichen sward forming a network within which patches are interspersed, either of bare ground with individual plants growing on it, or of ground which to a considerable extent is covered by a surface crust of crustose lichens.

This type of continuous vegetation was named a 'continuous polygonal network' (Figs. 11 and 12) by Matveyeva (1979, pp. 6, 13). The most important traits distinguishing this vegetation from that on spotted tundra consist of the insignificant participation of flowering plants and the complete lack of dwarf shrubs among them. According to Matveyeva (1979, p. 27), the cover of flowering plants within such vegetation on Mys Chelyuskin ranges from 3 up to 6–7%: '... the flowering plants do not form a closed cover but grow individually'. In the tundra phytocoenoses there is a much greater participation of flowering plants, which in general predominate in the above-ground as well as the below-ground phytomass, and a more definite intermingling of the root systems below the ground surface (Aleksandrova, 1977a). On Zemlya Aleksandra the cover of flowering plants within the continuous polygonal network amounts to 1–

Fig. 11. Polygonal network vegetation sward in a moss–lichen (*Cetraria delisei, C. islandica* var. *polaris, Rhacomitrium lanuginosum*) aggregation on a rocky silt-stone substrate with sorted material, on Mys Chelyuskin (photo.: N. V. Matveyeva).

6%, but the plants are very small and their root systems do not intermingle. In contrast to the situation in the tundra, the characteristic trait of this vegetation is the complete elimination of tiers (Aleksandrova, 1977a; Matveyeva, 1979; see also Figs. 13 and 31 in this book). It should be noted that the dimensions of the polygons in the arctic polar deserts are considerably smaller than those of the polygons in the arctic tundra. The reason for the smaller dimensions of the polygon formations in the polar desert has already been discussed.

The second type of closed vegetation cover occurring in the arctic polar deserts takes the form of sites covered by a superficial crust of lichens. It is met with in dry habitats and at sites where a cover of a black film of liverworts (here as individual structures), mucilaginous lichens and blue-green algae predominate in wet habitats. The participation of higher plants here is quite insignificant.

Fig. 12. Continuous polygonal network type of vegetation sward in a moss–lichen (*Cetraria delisei, Pertusaria glomerata, Ochrolechia frigida, Ditrichum flexicaule, Dicranoweisia crispula*) aggregation with participation of small tufts of *Phippsia algida* on structured ground with small basalt blocks, sorted out of fine soil, on the island of Zemlya Aleksandra (photo.: the author).

We are thus faced with the problem of deciding to which category of phytocoenotic units the polar desert vegetation with its continuous or discontinuous character belongs.

When examining sites with a polygonal network vegetation (Fig. 13), it can be seen that there are three elements of heterogeneity: a network of moss–lichen sward with a small mixture of flowering plants, a crust of predominantly crustose lichens and, between these, bare ground on which very small individual specimens of flowering plants and small patches of

Fig. 13. One of the best developed nanocomplexes on the island of Zemlya Aleksandra. Area of site shown: 1m². *1*, moss–lichen sward (predominantly *Cetraria cucullata, C. ericetorum* and *Ditrichum flexicaule*) together with small amounts of flowering plants (*Papaver polare, Cerastium arcticum, Saxifraga hyperborea, Phippsia algida*, etc); *2*, crust of crustaceous lichens (*Pertusaria spp., Ochrolechia* spp., *Collema ceraniscum*, etc.); *3*, bare loam with a small amount of stones, individual specimens of *Phippsia algida* and some small patches of lichens.

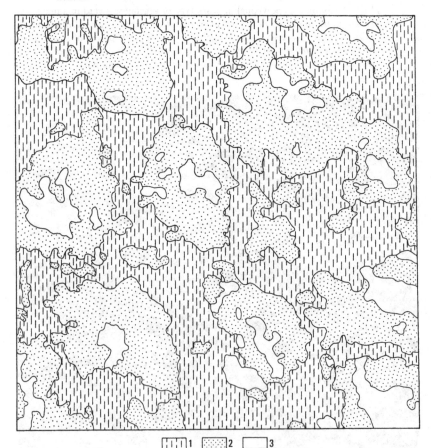

☐☐☐1   ☐☐☐2   ☐3

lichens and mosses occur. It is necessary to establish criteria to explain whether such a situation represents a locality with a single mosaic phytocoenoses or whether from the point of view of composition this heterogeneity should be considered to arise from 'nanocomplexes' which consist of two differing nanophytocoenoses, i.e., a crust of crustose lichens on the polygons with a network of moss–lichen sward surrounding them.

The most objective criteria for a solution to the problem of whether one is dealing with a mosaic or a complex have been suggested by Lavrenko (1959). According to Lavrenko, a mosaic occurs when the elements of heterogeneous more or less continuous communities are interconnected by the intermingling of the root systems of the plants throughout the entire assemblage. When distinct complexes are present, contact between plants belonging to different phytocoenoses or fragments thereof occurs only at borderline sites and the major parts of the components of the adjacent phytocoenoses do not make contact with each other.

These criteria were applied when studying the characteristics of the heterogeneity within the spotted tundras of the arctic (on Bol'shoy Lyakhovskiy) and the subarctic (in western Taymyr). In both cases, spots of naked ground on which different individuals of flowering plants, small patches of ground-living lichens, etc. grow are interspersed within a closed network of herb–dwarf bush–moss sward. In the composition of this sward many dwarf bushes are included: on the island of Bol'shoy Lyakhovskiy *Salix polaris*, in western Taymyr *Dryas punctata*; herbs are common: in the former locality, e.g., *Alopecurus alpinus* and *Luzula confusa*, in the latter locality, *Carex ensifolia* ssp. *arctisibirica*. When the root systems were dug up it could be seen that the underground organs (roots and rhizomes) of a number of sward-forming plants penetrated from under the sward into the open patches and completely permeated the soil below them. This formed the basis for assuming that the spotted tundra should not be considered as a complex but as a distinct mosaic phytocoenoses. In the subarctic tundra of western Taymyr a dense network is formed below the open spots by roots and rhizomes of *Carex ensifolia* ssp. *arctisibirica*, which grows on the mounds around the spots and thus unites the spotted tundra into a single phytocoenoses (Aleksandrova, 1971). In the arctic tundra on Bol'shoy Lyakhovskiy this role is fulfilled by the roots of *Salix polaris*, which intermingle with the rhizomes and roots of *Alopecurus alpinus* (Fig. 14).

A different aspect is presented by the vegetation of the arctic polar deserts. There the root systems of the plants do not come into contact with each other and, consequently, there is no connection between the elements belonging to an assemblage which could associate them into a single phytocoenosis (Fig. 15). The patches of sward, the superficial crusts of lichens and the interspersed spots of naked ground carrying individual plants are not interconnected and make contact only at the borders between them. Therefore we cannot consider the vegetation of the polar deserts to be an heterogeneous polygonal network vegetation forming mosaic phytocoenoses, but must instead consider it as different nano-complexes.

Plant associations like those illustrated in Fig. 13 can be considered to consist of nanocomplexes which are distinct from each other, i.e., repre-sent *nanophytocoenoses* forming a network of moss–lichen strips and crusts of crustose lichens on the polygons. The fact that these represent individual nanophytocoenoses is shown by the differences in the peculiar

Fig. 14. Vertical section through a mosaic photyocoenosis in the arctic tundra on the island of Bol'shoy Lyakhovskiy; the root system was dug out of the soil of a wall in a pit to a depth of 10 cm (according to Aleksandrova, 1962). *1, Salix polaris; 2, Alopecurus alpinus; 3, Draba pseudopilosa; 4, Potentilla hyparc-tica; 5,* unidentified roots; *6,* limits of root masses for a majority of roots; *7,* permafrost surface in August.

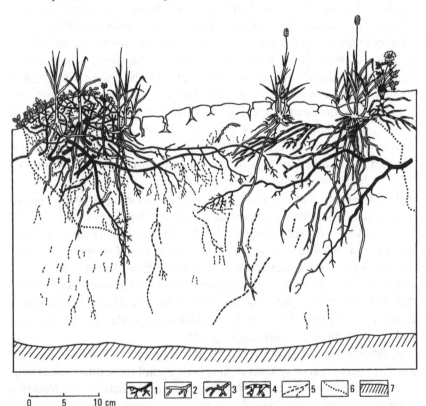

competitive relationship between the plants of each phytocoenosis. It is necessary to take note of the fact that it appears to be a specific characteristic of the polar deserts that the phytocoenotic relationship can be manifested within a very thin layer of 'a living film', in this case from 0.5 to 5 cm thick only. Observations have demonstrated that within the nanophytocoenoses represented by moss–lichen strips a constant struggle for space goes on: when developing, the plants suppress their neighbours, i.e., they crowd each other out. The least competitiveness occurs among the flowering plants but they are stifled by the mosses (for instance, mosses will grow through a clump of poppies as a result of which the poppies become overwhelmed and die) and, in their turn, crusts of crustose lichens and liverworts develop over the mosses as well as the flowering plants.

The participation of flowering plants in the nanophytocoenoses in question indicates that in place of the plants destroyed new ones will arrive (from seeds carried by the wind from sites where the flowering plants, especially the poppies, have an opportunity to produce ripe seeds, but mainly from the most favourable of such habitats, i.e., the 'stone hothouses') and gradually grow into juvenile specimens. Competition can also be found among the crusts of crustose lichens, which develop patches consisting of different species. Although the details of this kind of coexistence can be established only under the microscope, it does not diminish their role in the phytocoenosis. Lynge (1928, p. 70) stated, concerning similar interrelationships between crustose lichens on Novaya Zemlya: 'Frequently *Lecidea cyanea* (Ack.) Röhl. forces out other lichens;

Fig. 15. Vertical section through part of the best developed vegetation in the polar desert on Zemlya Aleksandra. *1*, mosses (*Ditrichum flexicaule*, etc.); *2*, crust of crustose lichens; *3*, lower, dead parts of mosses; *a*, bare ground; *P*, *Papaver polare*; *Ph*, *Phippsia algida*; *Ds*, *Draba subcapitata*; *S*, *Saxifraga hyperborea*; *Cc*, *Cetraria cucullata*.

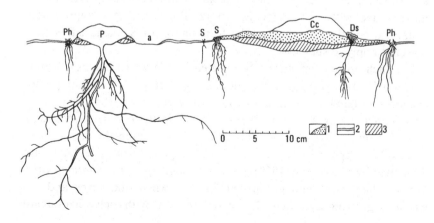

it appears to be a very "dangerous neighbour" for *Lecidea dicksonii* and a number of other lichens'. The mechanism of the competitive relationship between the lichens needs special research. A certain role is, of course, played by the different amounts of energy required for growth that are typical of individual species and also by the chemical effects of one species of lichen upon another.

A second example of phytocoenoses, i.e., the formation of plant environments, also takes place here. Plant environments form primarily within the limits of 'plant-producing areas' (Uranov, 1965) overlapping each other, within the limits of which unsatisfactory effects of adjacent species on each other also occur. In addition, the best developed nanophytocenoses of the polar deserts are able to transform the upper part of the substrates by making soils, the characteristics of which have been discussed above.

Nanocomplexes with a continuous plant cover have a very limited distribution within the arctic polar deserts since they are affiliated with narrow ecological niches having the most favourable habitat conditions. Open vegetation predominates and occupies wide territories.

Sukachev (1975) suggested in 1934 that vegetation where phytocoenotic contact between individual plants is completely absent should be called *aggregations*. The term aggregation has been used in this sense by Shennikov (1964) as well as by a number of other authors. However, in the polar deserts, besides aggregations in which no phytocoenotic contact between individuals exists, there are also found open associations where individuals, growing separately, form scattered phytochores within given limits and where there are distinct fragments, not touching each other, of nanophytocoenoses, the sizes of which range from a few square centimetres to a few square decimetres. Such phytochores cannot be considered as nanocomplexes because the fragments of the nanocomplexes do not make contact with each other (Figs. 16–18). If the concept of 'aggregation' is applied strictly according to the formulation of Sukachev, this term does not apply in such cases.

For such open aggregations, to the elements of which belong not only individual plants growing in isolation, but also fragments of nanophytocoenoses isolated from each other, I have suggested the term '*semi-aggregation*' (Aleksandrova, 1981). The majority of the associations identified on Zemlya Aleksandra but also found within the area of Mys Chelyuskin appear in the form of semi-aggregations. The latter are attributed by Matveyeva (1979) to *discontinuous types of vegetation sward*, growing like strips and cushions (Fig. 18), where the strips and the cushions, growing separately, do not make contact with each other but are isolated from each other by major or minor distances. According to the data of Matveyeva, these types of plant aggregations predominate in the area of Mys Chelyuskin.

Some of the fragments of nanophytocoenoses forming semi-aggregations are uniform in respect to their specific composition, i.e., some are represented by single synusium, while others are composed of two or more synusiae (see Fig. 18 and Figs. 34, 46 and 47 below).

The problem of how to classify the plant aggregations of the polar deserts in which the typological units are united arose for me in connection with 70 geobotanical relevés on Zemlya Aleksandra. For these relevés habitats were selected which showed comparatively uniform ecological conditions and which had a fairly uniform composition of plant species and range of characteristics within the limits of the ecotope in question. Thus, territorial units, i.e., phytochores, were described according to the concept of Norin (1970, 1979). The predominating portion of the phyto-

Fig. 16. Semi-aggregation of nanocomplex fragments and individual specimens of plants on polygonal loam with an admixture of small stones; on Zemlya Aleksandra. Dimensions of polygons: 8–15 cm broad and 11–20 cm long (photo.: the author).

chores is represented by open aggregations and only a small part of these have closed cover. The dimensions of the phytochores observed ranged from 10 m² or some multiples thereof up to 100 m².

In order to be able to carry out a classification of the phytochores, an attempt was made to utilize the similarity coefficient according to Jaccuard:

$$J = \frac{2c \cdot 100}{a + b},$$

(where a is the number of species of the first relevé, b the number of species of the second relevé and c the number of species common to both the first and the second relevé), as well as a formula according to B. I. Semkin, i.e.:

$$K_I(A, B) = \frac{\Sigma_{min}\,(p_i, q_i)}{\Sigma_{max}\,(p_i, q_i)}$$

(where A and B are the number of species in the phytochores being compared and p_i and q_i are the cover of the total number of species belonging to phytochores A and B, respectively).

However, the desired results could be only partially obtained for some of the best developed nanocomplexes, and for the open aggregations it was impossible to obtain adequate distinctions between aggregations because of a number of differences in the specific composition of similar ecotopes. It may very well be that this is due to difficulties in germination

Fig. 17. Semi-aggregation of nanocomplex fragments and individual specimens of plants on a ground cracked into polygons in the area of Mys Chelyuskin (photo.: N. V. Matveyeva).

and survival of the young plants (especially of cryptogamous plants) under these extremely severe conditions, as a consequence of which the composition of the species will to a considerable extent be dependent on factors of chance.

In order to elucidate how to group the phytochores into some primary units, an attempt was made at an ordination while associating the relevés with the following environmental factors: (1) the duration of the growing period and the total sum of positive temperatures occurring from the date when the snow disappeared; (2) the quality of the ground; and (3) the amount of moisture available (Fig. 19).

The procedure for classification was executed in three stages.

First stage: selection of preliminary groups of relevés. For this purpose the numbers of those phytochores with particular ecological characteristics were written within each of the cells in the ordination diagram. Their

Fig. 18. Example of semi-aggregation on surface measuring 1 m² on Mys Chelyuskin (according to N. V. Matveyeva, 1979). *1*, bare ground; *2*, *Orthothecium chryseum*; *3*, *Bryum tortifolium*; *4*, *Stereocaulon rivulorum*; *5*, *Thamnolia subuliformis*; *6*, *Phippsia algida*; *7*, *Cerastium regelii*; *8*, *Saxifraga oppositifolia*; *9*, *S. cernua*; *10*, fissures in the ground.

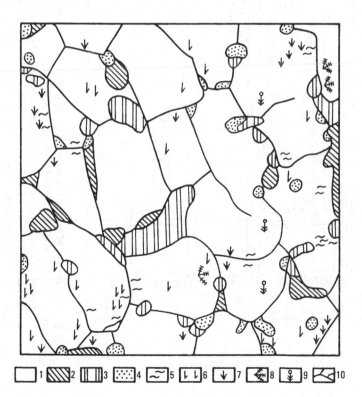

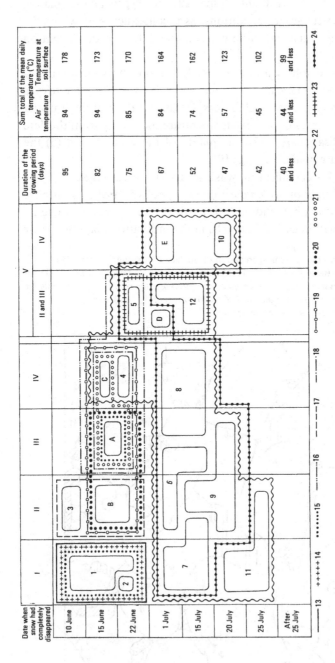

Date when snow had completely disappeared	I	II	III	IV	V	II and III	IV	Duration of the growing period (days)	Sum total of the mean daily temperature (°C) Air temperature	Sum total of the mean daily temperature (°C) Temperature at soil surface
10 June								95	94	178
15 June								82	94	173
22 June								75	85	170
1 July								67	84	164
15 July								52	74	162
20 July								47	57	123
25 July								42	45	102
After 25 July								40 and less	44 and less	99 and less

13 ———— 14 ++++ 15 ●●●●●● 16 ——— 17 – – – – 18 ——–– 19 o–o–o 20 ●●●●● 21 ooooo 22 ~~~~ 23 ++++++ 24 ●—●—●

Fig. 19. Ordination of ecological groups (eco-groups), types of nanocomplexes and types of open aggregations on Zemlya Aleksandra. *I*, rock heaps; *II*, sandy-gravelly-pebbly and sandy-gravelly-fine mineral soil; *III*, stones in combination with fine soil, in part including loam; *IV*, loam broken up into polygons with a little gravel, pieces of rock or pebbles included; *V*, increased moisture content. *A–E*, types of nanocomplexes (see pp. 62–85); *1–12*, types of open aggregations (see pp. 85–120). Eco-groups: *13*, *Umbilicaria*; *14*, *Rhizocarpon*; *15*, *Rhacomitrium lanuginosum*; *16*, *Rinodina–Ochrolechia*; *17*, *Stereocaulon vesuvianum* var. *depressum*; *18*, *Pertusaria–Ochrolechia–Collema*; *19*, *Ditrichum flexicaule*; *20*, *Cetraria cucullata*; *21*, *Hylocomium alaskanum*; *22*, *Stereocaulon rivulorum*; *23*, *Orthothecium chryseum*; *24*, 'black film'. The dates of the disappearance of the snow, the duration of the growing period and the sum total of the temperatures were recorded during 1959.

habitat conditions corresponded to a combination of special ecological characteristics. In this manner phytochores were united into preliminary groups on the basis of the similarity of their ecology.

Second stage: identification of the dominant ecological groups of species. For each of the preliminary groups of phytochores comparable lists of their specific composition were established. A scrutiny of these lists revealed the presence of adequately stenotopic or hemistenotopic combinations of species, which were met with in relation to a definite combination of ecological factors. Thus ecological groups of species were identified. Twelve such 'eco-groups' were established (Table 5).

The *Umbilicaria* eco-group includes epilithic, foliose lichens (*Umbilicaria* spp., mainly *U. proboscidea*, frequently *U. arctica* and occasionally individual specimens of *U. hyperborea, U. decussata* or *U. cylindrica*) and fruticose lichens (*Usnea sulphurea, Alectoria pubescens,* etc.) as well as the epilithic moss *Andreaea rupestris*. All are black or almost black except for *Usnea sulphurea*, which has a yellowish-green colour.

The *Rhizocarpon* eco-group consists of epilithic, crustose lichens. *Rhizocarpon geographicum* colours the surface of the stones in greyish or greenish-yellow tones, *Lecidea dicksonii* has a characteristic ochraceous colour, *L. macrocarpa* can be grey, white or greyish-brown, *Lecanora polytropa* greenish or yellowish-grey. A number of other species, described by A. N. Oksner on the basis of my collection, are also epilithic lichens. These are reported in the lists of species given below, e.g. Table 10.

The *Rhacomitrium lanuginosum* eco-group. Fruticose lichens such as *Alectoria ochroleuca, A. nigricans, Cornicularia divergens* usually occur together with the moss *Rhacomitrium lanuginosum*.

The *Rinodina–Ochrolechia* eco-group comprises crustose lichens met with on remnants of plants; these are the black, blackish or brownish-black *Rinodina turfacea, Pannaria pezizoides, Psoroma hypnorum* and the whitish *Ochrolechia frigida*.

The *Stereocaulon vesuvianum* var. *depressum* eco-group includes, in addition to the ground-hugging arctic–alpine lichen *Stereocaulon vesuvianum* var. *depressum*, a number of species which form the basis of the grey plant crust often found on the surface of sites with a scanty snow cover or on sandy-pebbly and sandy-gravelly soils. These species include a number of crustose lichens, e.g., *Lecanora campestris, Buellia caniops,* and also species capable of existing both on plant remnants and on fine soil, e.g., *Buellia punctata, Caloplaca jungermanniae* and other speces mentioned in Tables 8 and 12. The crust contains the small arctic–alpine moss *Encalypta alpina* and some other small mosses (e.g. *E. rhabdocarpa*, species of *Bryum, Pohlia*).

Table 5. *Ecological species groups on the island of Zemlya Aleksandra*

Eco-group	Species (typical representatives)	Distribution type
Umbilicaria	*Umbilicaria proboscidea* (L.) Schrad.	Arctic–alpine
	U. arctica (Ach.) Nyl.	Arctic–alpine
	Usnea sulphurea (Koenig) Th. Fr.	Bipolar arctic–antarctic
	Alectoria pubescens (L.) Howe	Bipolar arctic–antarctic
	A. minuscula (Nyl.) Degel.	Bipolar arctic–antarctic
	Cetraria hepatizon (Ach.) Vain.	Arctic–alpine
	Hypogymnia intestiformis (Vill.) Räs.	Arctic–alpine
	Andreaea rupestris Hedw. var. *acuminata* (B.S.G.) Sharp	Arctic–alpine
Rhizocarpon	*Rhizocarpon geographicum* (L.) DC.	Bipolar, multizonal; basically arctic-alpine and antarctic
	Lecidea dicksonii (Gmel.) Ach.	Arctic–alpine
	L. macrocarpa (DC.) Steud.	?Arctic–alpine
	Lecanora polytropa (Ehrh.) Rabenh.	Bipolar arctic–alpine and antarctic
Rhacomitrium lanuginosum	*Rhacomitrium lanuginosum* (Hedw.) Brid.	Bipolar, multizonal, almost cosmopolitan, widely distributed in the arctic
	Alectoria nigricans (Ach.) Nyl.	Bipolar arctic–alpine and antarctic
	A. ochroleuca (Hoffm.) Mass.	Arctic–alpine
	Cornicularia divergens Ach.	Arctic–alpine
	Sphaerophorus globosus (Huds.) Vain.	Bipolar, almost cosmopolitan, arctic–alpine and antarctic

Rinodina–Ochrolechia	*Rinodina turfacea* (Ach.) Koerb.	Bipolar arctic–alpine and antarctic
	Pannaria pezizoides (Web.) Trevis	Arctic–alpine
	Psoroma hypnorum (Vahl.) S. Gray	Bipolar arctic–alpine and antarctic
	Ochrolechia frigida (Sw.) Lynge	Bipolar arctic–alpine and antarctic
Stereocaulon vesuvianum var. depressum	*Stereocaulon vesuvianum* Pers. var. *depressum* (H. Magn.) M. Lamb	Arctic–alpine
	Lecanora campestris (Schaer.) Hue	Arctic–alpine
	Buellia punctata (Hoffm.) Mass.	Bipolar, cosmopolitan
	B. coniops (Whalenb.) The. Fr.	Bipolar, arctic–alpine and antarctic
	Caloplaca jungermanniae (Vahl.) Th. Fr.	Arctic–alpine
	Encalypta alpina Sm.	Arctic–alpine
Pertusaria– Ochrolechia– Collema	*Pertusaria glomerata* (Ach.) Schaer.	Arctic–alpine
	Ochrolechia frigida (Sw.) Lynge	Bipolar arctic–alpine and Antarctic
	Collema ceraniscum Nyl.	Arctic–alpine
Ditrichum flexicaule	*Ditrichum flexicaule* (Schwaegr.) Hampe	Arctic–alpine
	Polytrichum alpinum Hedw.	Bipolar arctic–alpine and antarctic
	Drepanocladus uncinatus (Hedw.) Warnst.	Bipolar arctic–alpine and antarctic, almost cosmopolitan
	Distichium capillaceum (Hedw.) B.S.G.	Bipolar arctic–alpine and antartic, almost cosmopolitan
	Bartramia ithyphylla Brid.	Arctic–alpine
Cetraria cucullata	*Cetraria cucullata* (Bell.) Ach.	Arctic–alpine
	C. nivalis (L.) Ach.	Arctic–alpine
	C. ericetorum Opiz	Arctic–alpine

Eco-group	Species (typical representatives)	Distribution type
Hylocomium alaskanum	*Hylocomium splendens* (Hedw.) B.S.G. var. *alaskanum* [Lesq. et James] Limpr. [*H. alaskanum* [Lessq. et James] Kindb.]	Arctic (an arctic taxon of a hyparctic species)
	Aulacomnium turgidum (Wahlenb.) Schwaegr.	Arctic–alpine
	Tomenthypnum nitens (Hedw.) Loeske var. *involutum* (Limpr.) C. Jens.	Arctic (an arctic taxon of a hyparctic species)
Stereocaulon rivulorum	*Stereocaulon rivulorum* H. Magn.	Arctic–alpine (?bipolar)[a]
	Polytrichum fragile Bryhn	Arctic
	Hygrohypnum polare (Lindb.) Broth.	Arctic–alpine
	Pogonatum urnigerum (Hedw.) P. Beauv. var. *subintegrifolium* (Arn. et Jens.) Moell.	Arctic–alpine (an arctic-alpine taxon of a multizonal species)
	Psilopilum cavifolium (Wils.) Hag. f. *nanum* Hag.	Arctic
	Seligera polaris Berggr.	Arctic
	Pohlia obtusifolia (Brid.) L. Koch	Arctic–alpine
	Bryum tortifolium Brid.	Arctic–alpine
Orthothecium chryseum	*Orthothecium chryseum* (Schwaegr.) B.S.G.	Arctic–alpine
	Bryum arcticum (R. Br.) B.S.G.	Arctic–alpine
	B. tortifolium Brid.	Arctic–alpine
	Campylium stellatum (Hedw.) Lange et C. Jens.	Multizonal (widely distributed in the Arctic)
	Oncophorus virens (Hedw.) Brid.	Arctic–alpine
'Black film'	*Cephaloziella arctica* Bryhn et Douin	Arctic–alpine
	Lophozia alpestris (Schleich.) Evans	Arctic–alpine
	Collema ceraniscum Nyl.	Arctic–alpine

Notes: [a] Lindsay (1975, p. 51) reported *Stereocaulon* cf. *rivulorum* H. Magn. from South Georgia.

The *Pertusaria–Ochrolechia–Collema* eco-group forms the basis of the variegated crust of epilithic lichens which develops under conditions of moderate moisture on a loamy surface. *Collema ceraniscum* forms black spots, *Pertusaria glomerata* white ones and *Ochrolechia frigida* white spots often with a yellowish tint. To this group also belong other species of *Pertusaria* (e.g. *P. freyii*), *Ochrolechia tartarea*, which forms bright yellow patches, and some epilithic mosses. *Ochrolechia frigida* is a species that is rather widely distributed on the island; it appears as a member of almost all the eco-groups and is included in the Rinodina–Ochrolechia group, as characterized above, as well. It grows both on soils and on plant remnants.

The *Ditrichum flexicaule* eco-group comprises mosses, more characteristic of an optimal development of the plant associations.

The *Cetraria cucullata* eco-group includes fruticose lichens, e.g., the yellowish *Cetraria cucullata* and *C. nivalis* and the brownish *C. ericetorum*. Their abundance indicates the most favourable conditions for the development of the vegetation.

The *Hylocomium alaskanum* eco-group comprises three species of mosses (*H. alaskanum*, *Aulacomnium turgidum* and *Tomenthypnum nitens* var. *involutum*) which are met with only in the form of a slight mixture under the most favourable conditions; their presence demands not only an optimal snow regime but also the presence of loam although the latter occurs only in the form of a mixture.

The *Stereocaulon rivulorum* eco-group consists of greyish fruticose lichens with brownish, often abundant apothecia. It is widely distributed on Zemlya Aleksandra but is best developed where the snow disappears in the early part of July. It is also found where eco-groups characteristic of aggregations developed on sites where the snow disappears still later or where the most impoverished aggregations are found. In this eco-group are also found rare arctic species of mosses such as *Seligeria polaris* and *Psilopilum cavifolium* f. *nanum*.

The *Orthothecium chryseum* eco-group includes mosses that develop mainly under conditions of surplus moisture. The golden-rosy and shiny *Orthothecium chryseum* and the red-coloured species of *Bryum* provide the basic coloration of those plant associations where these mosses are relatively abundant.

The eco-group of species forming the 'black film' is met with mainly on wet soil and consists basically of liverworts (e.g. Table 5, Table 9 and relevé no. 68). Under such conditions the liverworts have developed special growth types in the form of shoots with tightly overlapping leaves of a dark colour and with cells which often have increased dimensions and a definite thickening of their walls; these form a crust or a film, completely adhering to the substrate (Ladyzhenskaya and Zhukova, 1971). *Collema ceraniscum* and microscopic blue-green algae also take part in the formation of the black film.

The amplitude of the different eco-groups which served as the basis for the final (third) stage of classification of the phytochores is also plotted on the ordination diagram (Fig. 19).

Third stage: unification of the phytochores into typical units. During the examination of the areas covered by the eco-groups it became clear that the ordination diagram could be subdivided by a series of ecological contours, each of which is characterized by a definite combination of eco-groups and unites a more or less large number of the preliminary group of 'relevés'. These relevés proved to be similar to each other in respect of the participation of this or that eco-group. Although in each of the relevés as a rule no particular eco-group is fully represented, since sometimes certain species, and at other times different species are absent, this characteristic can be taken as the basis for a classification of the phytochores. For instance, under type B (Fig. 19, Table 8) are united three relevés in which an important role is played by species from the *Cetraria cucullata* group in the composition of the eco-groups. *C. ericetorum* shares participation with *C. cucullata* in relevé no. 33 (both species cover up to 8% of the soil surface) but in relevé no. 30 *C. cucullata* has a cover amounting to only 1% and in relevé no. 20 it is in general absent at the same time as there is a 15% cover of *C. nivalis*, which also belongs to this eco-group. In addition, data on cover were established for all groups of species and structures of phytochores. A correlation between the three relevés mentioned also exists in respect of these characteristics, confirming that it is correct to unite them into one single type.

In some cases different ecological units were distinguished on the strength of the fact that although the participation of the eco-groups was similar important differences were found in the abundance of particular species. For instance, type 2 (Fig. 19, Table 11) was distinguished on the basis of a dominant occurrence of *Rhacomitrium lanuginosum* (cover 40%), while at the same time in all the phytochores united under type 1, the mosses were in general few and *R. lanuginosum* met with in insignificant amounts only (Table 10). Within the limits of one of the major ecological contours (where just two eco-groups dominate, i.e., the black film and the *Stereocaulon rivulorum* eco-groups), four typological units (nos. 6–9) were distinguished: type no. 6 (Table 15) with a considerable abundance of *Stereocaulon rivulorum* (cover: up to 50%); type no. 7 (Table 16) with a generally minor cover of plants (5–12%) and with a cover of *Stereocaulon rivulorum* amounting to only 1–2%; type no. 8 (Table 17), which was distinguished according to the noticeable occurrence of *Phippsia algida*, which in a number of cases forms individual miniature cushions; and, finally, type no. 9, which consisted of very impoverished open aggregations where more than one half of the general cover was represented by fragments of black film (which seems to be clearly distinct from that belonging to type no. 7).

As a result of the classification carried out, the phytochores on Zemlya Aleksandra, represented by nanocomplexes, could be united into five types and the phytochores, represented by open aggregations, could be distinguished into twelve types.

The classification of the open vegetation groups is a real problem (Aleksandrova, 1981) in all those areas where they represent the appearance of climatic or edaphic climaxes. This also applies to the classification of, e.g., the open aggregations of vegetation in very dry deserts.

I was faced with the problem of what to name the units obtained through the classification of the open plant associations. For these, designation by the term 'association' such as is used for the classification of phytocoenoses did not seem suitable because of the lack of a tendency towards forming associations within the open groups.

Consequently, I used the expression 'types of open aggregations' for the designation of the units comprising similar open groupings. However, it is perhaps sensible to apply a shorter term, e.g., *comitium* (from the Latin verb *comitare*, to assemble). This term emphasizes the fact that the elements of the open aggregations (i.e., the individual plants of aggregations and the fragments of nanocomplexes within semi-aggregations) do not touch and cannot interact with each other and so only 'keep company' with each other, since they are isolated from each other by minor or major distances.

For this purpose I have employed the typological units distinguished (i.e., the types of nanocomplexes and the types of open aggregations) for describing the vegetation on Zemlya Aleksandra (see pp. 53–121). An attempt was also made to apply them for comparative analyses of the data concerning the vegetation on Mys Chelyuskin published by Matveyeva (1979).

The vegetation within the Barents province of the Soviet polar deserts

The Barents province of the polar deserts (see Fig. 1) is an area distinguished by a maritime, cryo-humid climate which covers Zemlya

Frantsa-Iosifa, the northern tip of Novaya Zemlya, Ostrov Victoria and, outside the USSR, the Nordaustlandet of the Svalbard archipelago (Aleksandrova, 1977a).

The basic difference between the vegetation of the arctic polar deserts belonging to the Barents province and that of those of the Siberian province consists of different associations which are characteristic of a given type of vegetation, e.g., the synusia of crustose lichens forming a crust on the soil and playing a major role in the composition of the plant aggregations. In the lichen crust on Zemlya Frantsa-Iosifa species of *Pertusaria* and *Ochrolechia* predominate. The combination of the white and the slightly yellowish-coloured patches of these lichens forms the 'background' synusia; in it black patches, formed by crusts of liverworts and lichens together with blue-green algal components, are distributed. On Mys Chelyuskin in the Siberian province (according to Matveyeva, 1979), the basic component in the formation of the crust of crustose lichens – which is of a greyish colour – is *Toninia lobulata*, which does not occur on Zemlya Frantsa-Iosifa. Korotkevich (1958) reminded us also of the grey lichens forming a crust on the soil when he described the vegetation of Severnaya Zemlya. Species of *Pertusaria*, which play a major role in the plant aggregations on Zemlya Frantsa-Iosifa, play practically no part in the formation of the ground crust on Mys Chelyuskin; in the lists of lichens from this area a few species belonging to the genus *Pertusaria* (*P. bryontha, P. bryophoga, P. dactylina, P. oculata* and *P. subdactylina*) are mentioned, but all with the remark 'very rare' (Piyn, 1979).

It is also necessary to note that in connection with the high humidity of the air over the Barents province and the frequent precipitation there, which favour the development of lichens, the lichen crust is of considerably greater importance in the formation of the plant aggregations than it is within the Siberian province, where the climate shows the characteristics of a more distinct continentality (although the lichen groups are also represented by synusiae characteristic of a polar desert type of vegetation).

A number of different species occur together although they have different geographical distribution areas. In addition to the example already mentioned of *Pertusaria* spp. playing an important role in the formation of the plant aggregations on Zemlya Frantsa-Iosifa (especially *P. glomerata*), but not being included in the lists of lichens from Mys Chelyuskin published by Piyn (1979), we can mention another species, very common among the lichens on Zemlya Frantsa-Iosifa, i.e., the crustose lichen *Usnea sulphurea*, which seems to be completely absent in the Siberian province of the polar deserts (the distribution area of *U. sulphurea* is mentioned below on p. 130).

Among the flowering plants on Zemlya Frantsa-Iosifa, apart from absolutely predominant circumpolar species such as *Phippsia algida* (Fig. 20), there are species which are not met with in the Siberian province, i.e., species with an amphiatlantic type of distribution. To these belong *Deschampsia alpina, Puccinellia vahliana, Cerastium arcticum* (Fig. 21), *Silene acaulis, Saxifraga caespitosa* ssp. *exaratoides,* etc. Among these *Cerastium arcticum* appears to be a constant component of many plant aggregations. Tolmachev (1931) considered the plants of this species distributed within high-arctic areas to be a distinct species, *Cerastium hyperborium* Tolm., although it should be mentioned that, as a consequence of this narrow opinion, 'the isolation of the high-arctic *C. hyperboreum* as a species, distinct from *C. arcticum* Lge. and narrowly

Fig. 20. Area of distribution of *Phippsia algida* (Soland) R. Br. (according to Hultén, 1968).

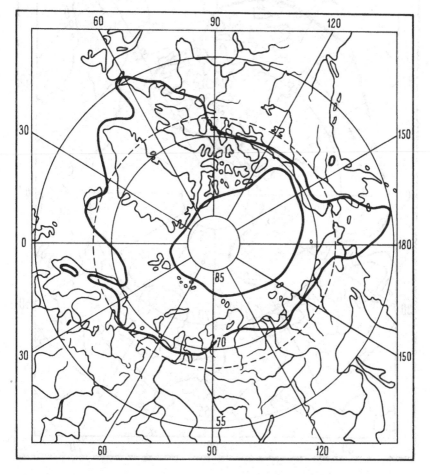

52   *The vegetation*

treated by us was not adequately founded ... Its distinction ... is not
sufficiently stable and not so important that it can serve as a basis for
maintaining specific independence of the high-arctic forms' (Tolmachev,
1971, p. 39). However, Böcher (1977) critically scrutinized all material
known up to then and came to the conclusion that the high-arctic form of

Fig. 21. Area of distribution of *Cerastium arcticum* Lange (according to data
from Porsild, 1957; Hultén, 1958; Böcher, Holmen and Jakobsen, 1968; *Arctic
flora of the USSR*, vol. 6, 1971).

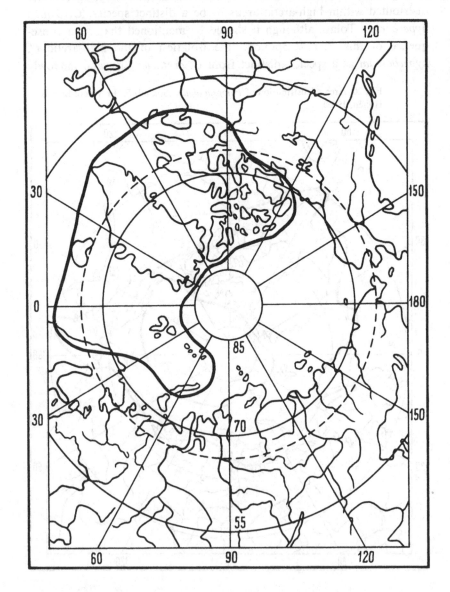

C. arcticum really represented a special taxon but of subspecific rank. He gave it the name *C. arcticum* Lge. ssp. *hyperboreum* (Tolm.) Böcher. According to Böcher, *C. arcticum* ssp. *hyperboreum* is well distinguished ecologically and geographically and has a typically arctic distribution from north-eastern America and Greenland all the way to Svalbard and Novaya Zemlya. It is not found in Scandinavia (Böcher, 1977). Accordingly, *C. arcticum* on Zemlya Frantsa-Iosifa should, following the opinion of Böcher, belong to *C. arcticum* ssp. *hyperboreum*. But according to Löve and Löve (1975) these taxa are not biologically conspecific since *C. arcticum* has $2n = 108$ chromosomes and *C. hyperboreum* has $2n = 54$.

Silene acaulis is met with only occasionally on the archipelago. Its distribution stretches across the Atlantic far westward, coming to a halt in north-eastern Asia (Fig. 22). *Saxifraga caespitosa* ssp. *exaratoides* was identified by Tolmachev from a specimen collected by Shukhtina from Ostrov Kheysa (Tolmachev and Shukhtina, 1974); according to Porsild (1957) it grows on the eastern islands of the Canadian Arctic Archipelago (it has not been mentioned as being found on Greenland).

The Siberian and the Siberian-American species which occur in the Siberian province of the polar deserts (Table 6) do not take part in the composition of the plant aggregations of the Barents province.

In order to evaluate the composition of the plant aggregations within the Barents province of the USSR (outside the USSR the north-eastern islands of the Svalbard Archipelago belong also to this province), I present below the material collected by myself on Zemlya Aleksandra. I give an account of these data while confining myself to those typological units whose basic characteristics are presented below.

Concerning the vegetation of other parts of the same province, e.g., the northern tip of Novaya Zemlya, there is only scanty information available, although what there is is very detailed (see pp. 121 and 124).

Zemlya Frantsa-Iosifa

Information about the vegetation of the archipelago of Zemlya Frantsa-Iosifa has been published by Esipov (1933), Leonov (1953), Govorukha (1960, 1968, 1970*a*), Mikhaylov and Govorukha (1962), Korotkevich (1972) and Aleksandrova (1969, 1977*b*).

More detailed data have been published in floras of flowering plants (Fischer, 1896; Palibin, 1903; Tolmachev, 1931; Hanssen and Lid, 1932; Tikhomirov, 1948*b*; Tolmachev and Shukhtina, 1974; Govorukha, 1960; Abramova *et al.*, 1961; Novichkova-Ivanova, 1963; Ladyzhenskaya and Zhukova, 1971, 1972; Zhukova, 1972, 1973*b*, 1973*c*, 1973*d*).

At present 57 taxa (species and subspecies) of flowering plants are known from Zemlya Frantsa-Iosifa (see Table 6) together with 102 taxa of mosses, 55 of liverworts and 115 of lichens.

The composition of the plant aggregations was studied by myself on Zemlya Aleksandra during 1959. This is the most well known island in this archipelago (Fig. 23). Its surface is up to 75% covered by two ice sheets.

Fig. 22. Area of distribution of *Silene acaulis* (L.) Jacq. (according to Hultén, 1968; *Arctic flora of the USSR*, vol. 6, 1971).

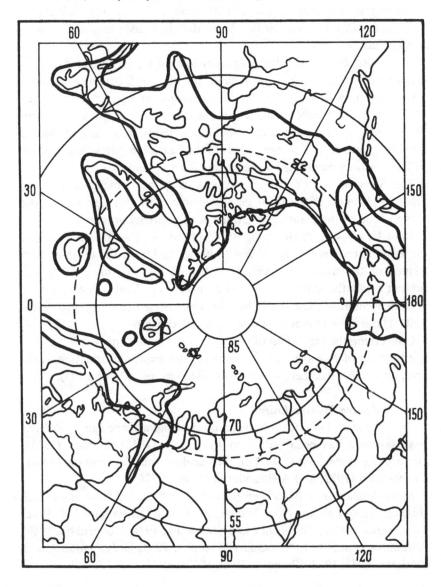

Table 6. *The flora of vascular plants of the Soviet arctic polar deserts*

Species, subspecies	Zemlya Frantsa-Iosifa		Severnaya Zemlya		Mys Chelyuskin area	Ostrov Bennetta	Distribution type
	All islands	Zemlya Aleksandra	Ostrova Komsomolets, Pioneer, Sedova, Diabazovyye	Ostrova Oktyabr'skoy Revolyutsii and Bol'shevik			
Alopecurus alpinus Sm.	+	−	+	+	+	+	Arctic–alpine, circumpolar
A. alpinus Sm. var. *borealis* (Trin.) Griseb.	−	−	−	+	−	−	Arctic–alpine, circumpolar
Arctagrostis latifolia (R. Br.) Griseb.	+	−	−	+	+	−	Arctic, circumpolar
Deschampsia alpina (L.) Roem. et Schult.	+	+	−	−	−	−	Arctic–alpine, amphiatlantic
D. caespitosa (L.) Beauv. ssp. *glauca* (Hartm.) Hartm.	+	+	+	+	+	−	Arctic–alpine, circumpolar
D. brevifolia R. Br.	−	−	−	+	−	−	High-arctic, circumpolar
Pleuropogon sabinii R. Br.	+	−	−	+	−	+	Arctic–alpine, circumpolar
Poa arctica R. Br.	+	−	−	+	+	−	Arctic, circumpolar
P. arctica R. Br. var. *vivipara* Hook.	+	−	−	−	+	−	Arctic, circumpolar
P. tolmatchewii Roshev.	−	−	−	−	−	−	Arctic–alpine, eurasiatic
P. lindebergii Tzvel.	+	−	−	+	+	−	Arctic–alpine, eurasiatic
P. alpigena (Fr.) Lindm.	+	+	+	+	+	−	Arctic–alpine, circumpolar
P. alpigena (Fr.) Lindm. var *colpoidea* (Th. Fries) Scholand.	−	−	+	+	+	−	Arctic–alpine, circumpolar
P. abbreviata R. Br.	+	+	−	+	+	−	Arctic, circumpolar
Dupontia fisheri R. Br.	+	−	−	+	+	−	Arctic, circumpolar

Table 6 (cont.)

Species, subspecies	Zemlya Frantsa-Iosifa		Severnaya Zemlya		Mys Chelyuskin area	Ostrov Bennetta	Distribution type
	All islands	Zemlya Aleksandra	Ostrova Komsomolets, Pioneer, Sedova, Diabazovyye	Ostrova Oktyabr'skoy Revolyutsii and Bol'shevik			
Phippsia algida (Soland.) R. Br.	+	+	+	+	+	+	Arctic, circumpolar
Ph. concinna (Th. Fries) Lindeb.	+	–	–	–	–	–	Arctic, eurasiatic
Puccinellia phryganodes (Trin.) Scribn. et Merr.	–	–	–	+	–	–	Arctic, circumpolar
P. angustata (R. Br.) Rand et Redf.	+	–	+	+	–	–	High-arctic, circumpolar
P. vahliana (Liebm.) Scribn. et Merr.	+	–	–	–	–	–	High-arctic, amphiatlantic
Festuca brachyphylla Schult.	–	–	–	+	–	–	Arctic-alpine, circumpolar
F. hyperborea Holmen	–	–	–	+	–	–	Arctic, E. Siberian–American
Eriophorum scheuchzeri Hoppe	–	–	–	–	+	–	Arctic-alpine, circumpolar
Carex stans Drej.	–	–	–	–	+	–	Arctic, circumpolar
C. ursina Dew.	+	–	–	–	–	–	Arctic, circumpolar
C. ensifolia (Turz. ex Gorodk.) V. Krecz. ssp. arctisibirica Yurtz.	–	–	–	+	–	–	Arctic, E. European–Siberian
Juncus biglumis L.	+	–	–	+	+	–	Arctic-alpine, circumpolar
Luzula confusa Lindb.	+	+	–	+	+	–	Arctic-alpine, circumpolar
L. nivalis Laest.	+	–	–	+	+	–	Arctic-alpine, circumpolar
Salix polaris Wahlenb.	+	–	–	+	+	+	Arctic-alpine, eurasiatic–W. American

Species						Distribution
S. arctica Pall.	+	−	+	+	+	Arctic–alpine, circumpolar
S. reptans Rupr.	−	−	+	+	+	Arctic, mainly Siberian
Oxyria digyna (L.) Hill	+	−	+	+	+	Arctic–alpine, circumpolar
Polygonum viviparum L.	+	−	−	+	+	Arctic–alpine, circumpolar
Stellaria edwardsii R. Br.	+	+	+	+	+	Arctic, circumpolar
S. ciliatosepala Trautv.	−	−	−	+	+	Arctic, circumpolar
S. crassipes Hult.	+	−	+	−	−	High-arctic, circumpolar
S. laeta Richards.	−	−	+	−	−	Arctic, circumpolar
S. humifusa Rottb.	−	−	+	−	−	Arctic, circumpolar
Cerastium arcticum Lge.	−	+	−	−	−	Arctic, amphiatlantic
C. regelii Ostenf. ssp. regelii	+	−	+	+	+	Arctic, circumpolar
C. regelii Ostenf. ssp. caespitosum (Malmgr.) Tolm.	+	+	+	+	+	High-arctic, mainly amphiatlantic but occurs also in arctic Siberia
C. beeringianum Cham. et Schlect. ssp. bialynickii (Tolm.) Tolm.	+	+	+	+	+	Arctic, Siberian–W. American
Sagina intermedia Fenzl	−	−	−	−	−	Arctic, circumpolar
Minuartia rubella (Wahlenb.) Hiern	+	+	+	−	−	Arctic–alpine, circumpolar
M. macrocarpa (Pursh) Ostenf.	−	−	−	+	+	Arctic–alpine, Siberian–W. American
Silene acaulis (L.) Jacq.	+	+	−	−	−	Arctic–alpine, European–American, occurs in E. Asia
Gastrolychnis apetala (L.) Tolm. et Kozh.	−	−	+	+	+	Arctic–alpine, circumpolar
Caltha arctica R. Br. ssp. arctica	−	+	+	+	+	Arctic, Siberian–W. American
Ranunculus hyperboreus Rottb.	−	−	−	−	−	Arctic, circumpolar
R. nivalis L.	−	−	+	+	+	Arctic, circumpolar
R. sulphureus Soland.	+	+	+	+	+	Arctic–alpine, circumpolar
R. sabinii R. Br.	+	+	+	+	+	High-arctic, circumpolar
Papaver polar (Tolm.) Perf.	+	−	+	+	+	High-arctic, circumpolar
P. radicatum Rottb. s.l.	−	+	+	+	+	Arctic, circumpolar
Eutrema edwardsii R. Br.	−	−	+	−	−	Arctic–alpine, circumpolar
Braya purpurascens (R. Br.) Bunge	−	−	−	+	+	Arctic, circumpolar
Cardamine bellidifolia L.	+	+	+	+	+	Arctic–alpine, circumpolar
Parrya nudicaulis (L.) Regel	−	−	+	−	−	Arctic–alpine, Siberian–W. American

Table 6 (cont.)

Species, subspecies	Zemlya Frantsa-Iosifa		Severnaya Zemlya		Mys Chelyuskin area	Ostrov Bennetta	Distribution type
	All islands	Zemlya Aleksandra	Ostrova Komsomolets, Pioneer, Sedova, Diabazovyye	Ostrova Oktyabr'skoy Revolyutsii and Bol'shevik			
Draba pilosa DC.	−	−	−	+	−	−	Arctic, Siberian–W. American
D. barbata Pohle	−	−	−	+	−	+	Arctic-alpine, E. Siberian
D. subcapitata Simmons	+	+	−	+	−	−	High-arctic, almost circumpolar
D. oblongata R. Br.	+	+	+	+	+	−	High-arctic, almost circumpolar
D. pauciflora R. Br.	+	+	+	+	+	−	Arctic-alpine, circumpolar
D. alpina L.	−	−	+	+	+	−	Arctic-alpine, circumpolar
D. kjellmanii Lid ex Ekman	−	−	−	+	+	−	High-arctic, European–Siberian
D. pohlei Tolm.	+	−	−	+	−	−	Arctic, Siberian
D. macrocarpa Adam	+	+	+	+	−	+	Arctic, circumpolar
D. lactea Adam	+	−	−	+	+	−	Arctic, circumpolar
D. pseudopilosa Pohle	+	−	−	+	−	−	Arctic-alpine, eurasiatic–W. American
Cochlearia arctica Schlecht.	+	−	−	+	+	+	Arctic, circumpolar
C. groenlandica L.	+	+	+	+	+	+	Arctic, circumpolar
Saxifraga nivalis L.	+	+	−	+	+	−	Arctic-alpine, circumpolar
S. tenuis (Wahlenb.) H. Smith	+	+	−	+	+	+	Arctic, circumpolar
S. foliolosa R. Br.	+	−	−	+	+	−	Arctic-alpine, circumpolar
S. hirculus L.	−	−	−	+	+	−	Arctic-alpine, circumpolar

Species	57	24	17	75	59	20	Distribution
S. platysepala (Trautv.) Tolm.	+	–	–	+	+	+	High-arctic, circumpolar
S. serpyllifolia Pursh	+	–	–	+	+	–	Arctic-alpine, E. Siberian–W. American
S. cernua L.	+	+	+	+	+	+	Arctic-alpine, circumpolar
S. caespitosa L.	+	+	+	+	+	+	Arctic-alpine, circumpolar
S. caespitosa L. ssp. *exaratoides* (Simm.) Engl. et Irmsch.	+	–	–	–	–	–	Arctic, amphiatlantic
S. hyperborea R. Br.	+	+	–	+	+	–	Arctic, circumpolar
S. oppositifolia L.	+	+	+	+	+	–	Arctic-alpine, circumpolar
S. rivularis L.	+	–	–	–	–	–	Arctic-alpine, circumpolar
Chrysosplenium alternifolium L.	–	–	–	+	+	–	Arctic-boreal, eurasiatic
Potentilla pulchella R. Br.	–	–	–	–	–	+	Arctic, circumpolar
P. hyparctica Malte	+	–	–	+	+	–	Arctic, circumpolar
Novosieversia glacialis (Adam) F. Bolle	–	–	–	+	–	+	Arctic-alpine, Siberian–W. American
Dryas octopetala L. s.l.	–	–	–	+	–	–	Arctic-alpine, almost circumpolar
D. punctata Juz.	–	–	–	+	+	–	Arctic-alpine, almost circumpolar
Androsace triflora Adam	–	–	–	+	+	–	Arctic, Siberian
Myosotis asiatica (Vesterg.) Schischk. et Sergo.	–	–	–	–	–	–	Arctic-alpine, eurasiatic–W. American
Eritrichum villosum (Ldb.) Bunge	–	–	–	+	+	–	Arctic-alpine, eurasiatic
Lagotis minor (Willd.) Standl.	–	–	–	+	+	–	Arctic-alpine, eurasiatic–W. American
Artemisia borealis Pall.	–	–	–	+	+	–	Hyparctic-montane, Siberian–American
Nardosmia frigida (L.) Hook.	–	–	–	–	–	+	Arctic-boreal, eurasiatic–W. American
Senecio atropurpureus (Ldb.) B. Fedtsch.	–	–	–	+	+	–	Arctic-alpine, Siberian–American
Saussurea tilesii (Ldb.) Ldb.	–	–	–	–	+	+	Arctic-alpine, E. Siberian
Total number of species	**57**	**24**	**17**	**75**	**59**	**20**	

Remarks: +, species present; –, species absent. The list has been composed using the following sources: for Zemlya Frantsa-Iosifa: Fischer (1896), Palibin (1903), Tolmachev (1931), Hanssen and Lid (1932), Tikhomirov (1948b), Tolmachev and Shukhtina (1974), Safronova (1983); for Severnaya Zemlya: Korotkevich (1958), Khodachek (1980), Safronova (1981b); for Mys Chelyuskin: Safronova (1979); for Ostrov Bennetta: Tolmachev (1959).

Included in the number of arctic–alpine species are also 'arctic–alpine tundra' species and 'meta-arctic' species *sensu* Yurtsev (1977).

The non-glaciated part of the island (Tsentral'naya Susha), amounting to about 150 km², is represented by a basaltic plate, inclined towards the north-west, where it passes below sandy-gravelly sediments belonging to a low (8–10 m high) marine terrace, which, according to Dibner (1965), was deposited 2500–5000 years ago.

In the centre of the Tsentral'naya Susha there is a plain at 20–25 m a.s.l. with ridges of basalt, in part covered by thin moraine deposits in the form of loams with rocks and boulders. Along its perimeter, on its border towards the marine terrace, extends an ancient sandy-pebbly coastal bank (Glavnyy Val, formed 6000–7000 yrs B.P.), which reaches an elevation of about 20 m a.s.l. It can be traced to a number of other islands in the archipelago as well (Dibner, 1965). In the southern part of the Tsentral'naya Susha, adjoining Zaliv Dezhneva, the ground (age 8000–10 000 yrs) rises up to 35 m a.s.l., in places to 45 m; the relief there is very broken and furrowed in the form of basaltic ridges with longitudinal lake basins in between.

The peculiarities of the climate arise from the Icelandic low pressure system and the cold transarctic marine current that brings ice with it from the central arctic basin. A low mean monthly summer air temperature (June: −1.3°C; July: 1°C, August: 0.5°C), a remarkably high winter temperature (−20.7°C), very strong winds and, for such a high latitude (81°N), a considerable amount of precipitation (*c.* 300 mm/a), almost exclusively in the form of snow, are characteristic of this area. The average

Fig. 23. Map of Zemlya Frantsa-Iosifa (according to Dibner, 1965). *1*, glaciers; *2*, areas free of ice.

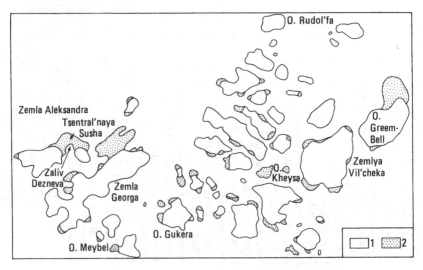

thickness of the snow cover at the end of the winter is about 60 cm; because of the strong local winds, the snow is rather unevenly distributed over the relief. The weather during the spring, summer and autumn is extraordinarily variable: not only is it cold but it is also damp, due to the almost daily precipitation (although only in small amounts at any one time), most often in the form of snow, rarely as a light rain or sleet; low evaporation; locally frequent fogs and a considerable humidity prevail as well.

In spite of the relatively small dimensions of the Tsentral'naya Susha, there are some differences in the weather of its northern part, which is washed by the waters of the Arctic Ocean with year-round ice floes occurring at the beaches, compared to that of the southern side, facing Zaliv Dezhneva. In the northern part of the island there are frequent fogs: at Zaliv Dezhneva there is less fog (the absence of open water in Zaliv Dezhneva plays a certain role; this bay is covered with ice all year round and narrow leads open along the coast only during the summer). The relief along Zaliv Dezhneva is very broken, resulting in more favourable microclimatic conditions at some sites. Therefore the thaw starts earlier in the coastal areas of Zaliv Dezhneva and the seasonal thawing of the permafrost occurs faster there.

The vegetation on Zemlya Aleksandra is extremely poor. On 25% of the surface of the Tsentral'naya Susha, where the snow lasts until 20 to 25 July, in general no plants are found. There are heaps of rocks below the slopes and on the lower parts of the slopes, stone fields in the depressions between the hills and sandy-loamy deluvium at the foot of some slopes and other localities where the snow reaches a depth of 1 m. About 35% of the land of the Tsentral'naya Susha is occupied by areas with uniform and very impoverished open aggregations, in which small individual plants, represented by a small number of species (< 10, including both higher and lower plants), hide between stones, are scattered at considerable distances from each other and cover a negligible percentage of the surface of such habitats. About another 20% of the ground is occupied by open aggregations, where the cover of the plants amounts to 3–15%. Only on barely 1% of the surface of the Tsentral'naya Susha is it possible to find localities with a continuous plant cover (i.e., nanocomplexes). Admittedly, 10% of the territory is occupied by lakes and permanent snow fields.

After working up all the material I had collected on Zemlya Aleksandra, the phytochores described were united into five types of nanocomplexes and twelve types of open aggregations. Their ecological ordination is illustrated in Fig. 19, but special lists of their specific composition are given below.

Types of Nanocomplexes

Nanocomplexes occupy only about 1% of the surface of the Tsentral'naya Susha on Zemlya Aleksandra. The plant cover is continuous but not unbroken: its continuity is interrupted by bare stones and interspersed patches of bare ground in which individual plants can be found.

During the investigation of the vegetation on this island special attention was given to the study of such aggregations, which could be called zonal, or representative of 'climatic climaxes', i.e., aggregations developed on plakor (Aleksandrova, 1980) habitats with a loamy soil under conditions of a moderate snow regime and a moderate atmospheric humidity. The appearance of such aggregations allows us to carry out a comparison between them and related aggregations in other areas under analogous habitat conditions in order to identify them and to discuss their position within a particular geobotanical region or belt.

I have described aggregations of such a kind as these from a plakor habitat (elevation *c.* 25 m a.s.l.) on polygonal loam where the snow was gone by 15 June. The plant cover was a continuous polygonal network and belongs, thus, to the category of nanocomplexes. The general plant cover amounts to 85%, of which 10–15% consist of a moss–lichen cover forming a network and 70–75% of a thin crust of predominantly crustose lichens covering the major portion of the soil polygons; the cover of the flowering plants was about 3% (see type C, Table 9). This aggregation can, due to its special structure and specific composition, be considered an arctic polar desert type of vegetation, the distinguishing characteristics of which have been summarized in detail by myself (Aleksandrova, 1969, 1977*a*) and are briefly repeated in this book in the concluding chapter (p. 196).

Nanocomplexes which are more satisfactorily developed than those of the zonal aggregations also occurred on sites which experience an optimal snow regime but which are on a substrate with a considerable amount of stones in the fine soil (nanocomplexes A and B). Stones jutting up above the plant cover have a warming effect on the ecotope (see p. 24). Although these nanocomplexes (aggregations of 'edaphic climaxes') are phytocoenotically richer than the zonal associations (they have a larger number of species and the strips of mosses and crustose lichens are wider and thicker), the peculiarities of their structure do not have tundra characteristics and they approach, like the zonal associations, typical polar desert vegetation: there are no dwarf shrubs, the flowering plants are small and few in number, their root systems do not intermingle and tiers are absent (Fig. 15). Such nanocomplexes have certain characteristics in common with the nanocomplexes typical of the southern belt of the polar deserts, as described by Matveyeva (1979) for Mys Chelyuskin, and because of this examples can be seen among them of extrazonality. The general cover of the plants amounted to 60–95%, of which 20–45% consisted of a sward of mosses and lichens, the rest being a crust on the soil, predominantly made up of lichens. The cover of the flowering plants amounted to 2–6% only (Table 7).

Nanocomplexes that were more impoverished than the zonal formations were met with on moist sites, where the snow disappears later (types D and E). There the continuity of the cover depends on the presence of a black film, which covers the substrate almost completely. The optimal cover of plants was 90–99%, mosses and lichens constituting 2–20% and the flowering plants (among which the presence of *Phippsia algida* tussocks is typical) 2–9%; the rest was occupied by the black film (see Table 9).

Since plant associations which are similarly developed and even more impoverished are described later, I begin the description of the nanocomplexes with types A and B, thereafter proceed to the zonal nanocomplexes (type C) and, finally, come to the D and E nanocomplexes.

Nanocomplexes of type A. These are aggregations with a polygonal network, consisting of mosses and fruticose lichens together with a considerable number of species from *Cetraria cucullata* eco-group (Table 7.)

Nanocomplexes of this type include the phytocoenotically richest plant aggregations, which are met with occasionally on Zemlya Aleksandra and which comprise the largest number of species. The highest number of eco-groups are also found in their composition. Nanocomplexes included in the type are associated with the most favourable conditions but are found only in small-size habitats, where the substrate consists of a mixture of stones and fine soil together with an amount of loam. The species belonging to the nanocomplexes which are grouped under subtype A″ are especially numerous. There, the cover of the flowering plants amounts in some cases to 7%, and species that are rare on the island, such as *Luzula confusa* and *Poa alpigena*, participate, although as single individuals only; the root systems of the flowering plants do not mingle (Fig. 15).

Subtype A′. Moss–lichen aggregations with a considerable participation of lichens from the *Cetraria cucullata* eco-group.

Relevé no. 7. A level habitat at *c.* 25 m elevation a.s.l., 3 km west of the Nagurskaya polar station. The snow regime is optimal: the snow had disappeared by 15 June. The nanocomplexes sampled occupy areas within

Table 7. *Floristic composition and indices of estimated plant cover (in %) within nanocomplexes of type A (subtypes A′ and A′′)*

Species	A′				A′′		
	7	9	13	15	18	28	45
Grasses							
Phippsia algida	+	1	1	+	+	+	+
Poa abbreviata	+	1	1	+	+	+	+
P. alpigena	−	−	−	−	+	+	+
Rushes							
Luzula confusa	−	−	−	−	−	−	+
Herbaceous plants							
Cardamine bellidifolia	2	1	+	6	6	2	3
Cerastium arcticum	+	−	−	+	+	+	+
C. regelii ssp. *caespitosum*	+	+	+	+	−	+	+
Cochlearia groenlandica	+	−	+	−	−	−	+
Draba macrocarpa	−	−	−	−	−	−	−
D. oblongata	+	+	+	+	+	+	+
D. pauciflora	+	+	−	+	−	−	+
D. subcapitata	+	−	+	+	−	+	+
Papaver polare	1	+	+	1	4	1	2
Saxifraga caespitosa	+	−	−	1	+	−	+
S. cernua	+	+	−	−	+	+	+
S. hyperborea	+	+	+	+	+	+	+
S. oppositifolia	−	+	+	+	−	−	+
Stellaria edwardsii	+	+	+	4	+	+	+
Mosses	25	20	10	20	25	20	20
Aulacomnium turgidum	+	+	−	+	2	+	+

Species	1	2	3	4	5	6	7
Bartramia ithyphylla	–	–	+	–	–	+	–
Bryoerythrophyllum ferruginascens	–	–	–	–	–	–	+
Bryum pseudocrispulum	+	1	–	+	+	+	–
B. pseudotriquetrum	–	–	–	–	–	–	+
B. rutilans	+	–	+	1	–	–	+
Dicranoweisia crispula	1	–	4	–	1	1	1
Dicranum elongatum	–	–	1	+	–	–	1
Distichium capillaceum	+	–	6	7	–	+	–
Ditrichum flexicaule	5	8	+	1	4	9	5
Drepanocladus uncinatus	1	–	1	–	+	1	–
Hylocomium alaskanum	+	–	1	–	–	3	1
Hypnum revolutum	+	1	–	–	–	–	+
Mnium blyttii	+	–	–	–	–	–	–
M. rugicum	–	+	–	–	–	–	–
Myurella julacea	–	–	–	–	+	+	–
Oncophorus compactus	+	–	–	1	–	–	–
Orthothecium strictum	–	–	–	7	–	–	–
Pohlia cruda	+	1	3	–	+	+	2
Polytrichum alpinum	2	6	–	3	1	1	4
Rhacomitrium canescens	1	–	1	1	3	1	–
Rh. lanuginosum	7	+	–	+	–	1	10
Schistidium strictum	–	–	+	–	–	–	–
Timmia austriaca var. *arctica*	1	2	–	1	–	+	–
Tomenthypnum nitens var. *involutum*	+	–	5	+	+	1	–
Tortula ruralis	1	–	–	1	–	1	–
Liverworts							
Aplozia crenulata	–	–	–	•	–	–	•
Blepharostoma trichophyllum	–	•	–	•	•	•	•
Cephaloziella arctica	•	•	•	•	•	•	•
C. hampeana	–	–	–	–	–	–	•
Lophozia alpestris	•	–	•	•	•	–	–

Table 7 (cont.)

Species	A'				A''		
	7	9	13	15	18	28	45
L. excisa	—	*	—	*	—	—	*
L. major	*	—	—	—	—	*	—
Orthocaulis atlanticus	*	—	—	—	—	—	—
Scapania calcicola	*	*	*	*	*	—	—
Sphenolobus minutus	*	*	—	*	*	*	—
Tritomaria heterophylla	*	*	—	—	*	—	*
T. quinquedentata	—	—	—	*	—	—	*
T. scitula	—	—	—	—	—	—	—
Fruticose lichens							
Alectoria jubata var. chalybeiformis	20	20	10	20	20	25	20
A. nigricans	1	+	—	—	+	—	+
A. nitidula	+	—	—	+	—	—	—
A. ochroleuca	+	—	—	+	+	+	1
A. pubescens	+	—	—	—	—	—	+
Cetraria cucullata	1	11	2	10	15	10	10
C. delisei	—	1	—	—	—	—	—
C. ericetorum	5	6	5	4	2	10	5
C. hepatizon	—	—	—	—	—	—	+
C. islandica var. polaris	8	1	2	1	2	—	—
C. nivalis	3	—	—	4	—	1	1
Cornicularia divergens	1	—	—	—	2	2	1
Sphaerophorus globosus	—	—	—	—	—	—	—
Stereocaulon alpinum	—	—	+	1	+	—	+
S. rivulorum	—	1	+	—	+	—	1
S. vesuvianum var. pseudofastigiatum	—	—	—	—	—	2	—

Species	1	2	3	4	5	6	7
Usnea sulphurea	+	–	–	–	+	–	–
Foliose lichens							
Parmelia omphalodes	+	2	1	+	1	+	+
Peltigera rufescens var. *incusa*	+	2	+	+	–	+	+
Physcia muscigena	–	–	+	+	–	+	–
Solorina bispora	–	–	–	–	–	+	–
Umbilicaria proboscidea	+	–	–	–	1	–	–
Tubiform lichens							
Cladonia pocillum (horizontal thallus)	+	–	–	–	–	–	–
C. pyxidata	+	+	–	+	–	+	+
Thamnolia vermicularis	+	+	+	+	–	+	+
Crustose lichens							
Caloplaca jungermanniae	–	–	–	–	–	●	●
C. stillistidiorum	–	–	–	●	–	●	–
Collema ceraniscum	●	(●)	(●)	–	●	●	●
Lecanora epibryon	–	–	–	–	–	●	–
Lecidea sp.	–	–	–	–	–	●	–
Lepraria sp.	–	–	–	●	–	●	–
Ochrolechia frigida	(●)	(●)	(●)	●	●	–	●
O. cf. *tartarea*	(●)	–	–	●	(●)	–	●
Pannaria pezizoides	(●)	–	●	●	–	–	–
Pertusaria sp.	●	●	●	●	●	●	●

Table 7 (cont.)

Psoroma hypnorum	—	*	—	*	—	—
Rinodina turfacea	—	*	—	—	*	(*)
Coverage of light-coloured crustose lichens	22	21	30	25	30	18
Cover of black crust of crustose lichens and liverworts	20	19	30	20	26	15
Total cover of plants	85	80	75	85	88	85	70

Remarks: Here as well as in Tables 8–21 the numbers above the columns indicate the reference number of the relevés; + indicates a cover of <1%; – indicates species absent; * indicates a species listed once in our samples; (*) indicates species not included (or omitted) in the samples but whose presence was ascertained by visual observation; the – between the asterisks indicates species not listed in our samples and not confirmed by visual observation but whose absence cannot be proven. The general cover of crustose lichens and liverworts is not listed because under field conditions it was not possible to differentiate the black spots formed by crustose lichens and the blue-green algal phycobionts belonging to the families Collemataceae and Pannariaceae from black crusts of aggregations with liverworts. The cover for all of the black crust is therefore given instead of separate cover values for them and the light-coloured crustose lichens which form white, pale yellow and greyish spots. Because there is no assurance that all the species of crustose lichens and liverworts growing in these localities have been reported, the lists of crustose lichens and liverworts must be considered incomplete.

The mosses and the liverworts were identified by A.L. Abramova, A.L. Zhukova and R.N. Shlyakov and the lichens by A.N. Oksner, N.S. Golubkova and M. Lamb.

a stone net, where the substrate consists of a mixture of stones and fine soil; the dimensions of these 'medallions' average 2 × 3 m. The largest stones, which are covered by a small amount of epilithic lichens, jut up above the level surface of the medallions and occupy 10–15% of the surface of the ground. About 1–2% of the ground consists of small, loamy patches with the characteristics of cryogenic activity. The vegetation, covering 85% of the ground, is distinguished by a brownish-yellow coloration: the brownish tinges come from mosses, the yellowish colour seen everywhere on the mosses is due to the pale yellow lacework of *Cetraria nivalis*. There are many white and black crusts of crustose lichens and liverworts. At first glance the flowering plants seem quite imperceptible: they are small, usually buried in the moss–lichen sward and scattered about.

Relevé no. 9. Same area. A level habitat, its substrate loamy with a mixture of rocks. The snow was gone by 15 June. The nanocomplex has a polygonal structure (Figs. 24 and 25). The general cover of the plants is 80%. The major portion of the polygonal surfaces is covered by a mosaic of patches of crustose lichens with some liverworts, through which stones, c. 7 × 10 cm, jut up. Patches of bare ground with a mixture of gravel and small stones are also found. The polygons are surrounded by strips and rows of mosses and lichens (*Cetraria cucullata* and *C. ericetorum* predominate) mixed with small tufts of poppies. Moss–lichen strips fill the depressions in the cracks and reach 2–4 cm (rarely 6 cm) above the surface of the polygons. The strips are 3–20 cm wide. The majority of the poppy tufts grow buried among the mosses (see Fig. 15); during the period of my investigation almost all of them remained sterile, only a few starting to produce flowers by the end of the summer but even then the capsules never setting any seeds.

Relevé no. 13. Same area. A gentle slope with a northeasterly exposure. The snow was gone by 22 June. Medallions of fine soil (Fig. 26), measuring 1.5 × 2.5 m, consist of small stones (the stones jutting up above the soil surface occupy 15% of its surface) mixed with loam which, locally, forms small patches with traces of fibres (occupying 10% of the ground surface). The general cover of plants is 75%. The background consists of a superficial crust of white or black spots, a brownish-green mat of mosses, dark brown cushions of *Cetraria ericetorum* and yellowish patches of *C. cucullata* and *C. nivalis*. Flowering plants are almost imperceptible in the general aspect.

Relevé no. 15. Same area. A slight incline towards the south. The snow was gone by 15 June. Medallions of fine soil, 0.7–2 m broad and somewhat extended, cover the slope and are distinguished by brightly coloured yellowish flecks of *Cetraria cucullata* and greenish-brown patches of mosses. *Saxifraga oppositifolia* is relatively abundant (cover: 4%) but almost completely sterile. In the course of the summer of my investigation only five flowers appeared; the capsules did not open and the plants did not display any capsules from previous years. Black and dead shoots predominated within the tussocks.

I found considerably better developed specimens of *Saxifraga oppositifolia* on the eastern part of the shore along Zaliv Dezhneva, where, on 14

Fig. 24. One of the best developed nanocomplexes on Zemlya Aleksandra (Relevé no. 9, Table 7). Area of the plot: 1 m². *1*, moss–lichen sward (*Cetraria cucullata, C. ericetorum, Ditrichum flexicaule,* etc.); *2*, crust of mainly crustose lichens; *3*, bare loam with some mixture of stones; *4*, shoots of *Polytrichum alpinum*; *5, Thamnolia vermicularis*; *6, Phippsia algida*; *7, Papaver polare*; *8, Saxifraga hyperborea*; *9, S. cernua*; *10, Cerastium arcticum*; *11, C. regelii* ssp. *caespitosum*; *12, Draba oblongata* and *D. pauciflora*.

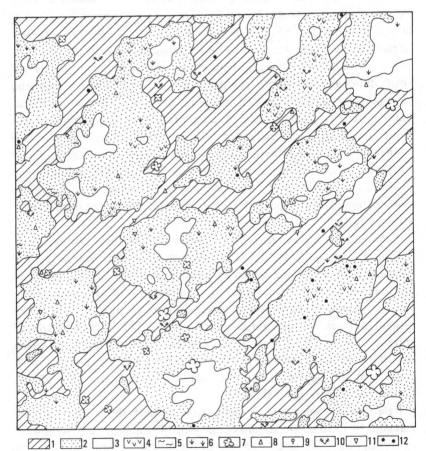

August they had still a relative abundance of flowers: there were many ripe capsules as well as capsules from the preceding year. The tufts were well developed, but dark green dead shoots almost completely covered the fresh living shoots and soft leaves. These habitats on the shore of Zaliv Dezhneva have a favourable relief, which protects them from the cold and wet winds blowing in from the ocean, in contrast to that of the habitats on the northern part of the island, where relevé no. 15 was situated.

The rest of the flowering plants, among which there are small dark green tussocks of poppies, did not attract attention. The general cover of plants amounted to 85%. About 50% of the substrate is occupied by a variegated black and yellow surface crust with some patches of loam as well as small stones jutting up above it.

Subtype A″. Moss–lichen aggregations with a considerable participation of the *Cetraria cucullata* eco-group and an increased number of flowering plant species.

Fig. 25. One of the best developed plant aggregations on Zemlya Aleksandra (Relevé no. 9, Table 7) (photo.: the author).

Relevé no. 18. A plakor habitat at a distance of 3 km west of the Nagurskaya polar station with an elevation of *c.* 25 m a.s.l. There is the beginning of a slight inclination towards the east. The snow was gone by 15 June. The substrate consists of small rocks mixed with loam. Fragments of rocks, measuring 4–10 cm (up to 15 × 20 cm), form a weakly outlined network on which a moss–lichen vegetation with *Cetraria cucullata* is distributed. The width of the moss–lichen strips is 3–12 cm; they are frequently interrupted by patches with a thin crust of crustose lichens. On the indistinctly formed polygons (width: 50–60 cm) a mosaic of crusts of crustose lichens together with some *Thamnolia vermicularis* has developed. Small patches of bare ground and stones, pushing up from below the sward, can also be seen. The general plant cover amounts to 88%. There are relatively many flowering plants (cover: 6%) but they are barely noticeable in the general aspect. Poppies predominated among the flowering plants; during my investigation the major portion of their small

Fig. 26. The vegetation on a medallion of fine soil surrounded by a stone net and with an optimal snow regime (photo.: the author).

Fig. 27. Fragments of vegetation on a stone net (Relevé 29, *comitium* no. 1; Table 10) with medallions of fine soil (Relevé no. 28, a nanocomplex of type A; Table 7), on Zemlya Aleksandra. The snow was gone by June 15th.

Vegetation on the stone net habitat (Relevé 29): *1, Umbilicaria proboscidea; 2, Alectoria pubescens; 3, Usnea sulphurea; 4, Andreaea rupestris; 5,* epilithic crustose lichens (*Lecidea macrocarpa, L. dicksonii,* etc.); *6,* crustose lichens on remnants of plants (*Ochrolechia frigida, Pannaria pezizoides,* etc) together with a small number of fruticose lichens (*Sphaerophorus globosus, Alectoria nigricans,* etc.); *7,* mosses (*Rhacomitrium lanuginosum, Dicranoweisia crispula*); *8, Stereocaulon vesuvianum* var. *pseudofastigiatum; 9,* bare stones.

Vegetation on the medallions of loam mixed with rocks (Relevé no. 28): *10, Thamnolia vermicularis; 11,* mosses (*Ditrichum flexicaule, Polytrichum alpinum, Pohlia cruda, Bryum pseudocrispulum, Aulacomnium turgidum,* etc.); *12,* fruticose lichens (predominantly *Cetraria cucullata* and *C. ericetorum*); *13,* crust of crustose lichens (*Pertusaria* spp., *Ochrolechia frigida, Rinodina turfacea, Collema ceraniscum,* etc.) together with some liverworts (*Cephaloziella arctica,* etc.); *14, Phippsia algida; 15, Polytrichum alpinum* (individual shoots); *16,* gravel between heaps of stones (to the left) and fine soil medallions (to the right); *P, Papaver polare; D, Draba oblongata; C, Cerastium arcticum; S, Saxifraga hyperborea.*

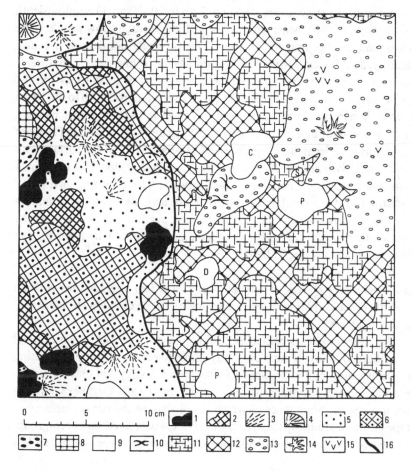

cushions remained sterile although some started to flower but did not set any seeds.

Relevé no. 28. Same area. Barely noticeable inclination towards the south. The snow was gone by 15 June. The medallions of fine soil (a mixture of small stones and loam), which were studied, have a width of 2 m and a length of up to 3.5 m. They are distinguished by pale yellowish cushions of *Cetraria cucullata* against a greenish-brown background of mosses. Among them there are scattered patches of white 'Thamnolia worms', black, slender and loosely branched specimens of *Alectoria nigricans* and *Cornicularia divergens*, a few cushions of *Cetraria ericetorum* and some patches of bare ground; dark green tufts of poppies can occasionally be seen but the rest of the flowering plants do not catch the eye easily (Fig. 27). The general plant cover amounted to 85%.

Relevé no. 45. The upper portion of a gently sloping hill, to the south, at *c*. 30 m elevation a.s.l. on the right-hand shore of Zaliv Dezhneva. The thickness of the snow cover is from 6–7 up to 20–25 cm. (The snow depth of the habitats was measured at sites with an average snow profile.) The snow had disappeared by 15 June. The substrate consists of a coarse disintegrated basalt, covered with a thin layer of fine soil (a mixture of loam and basaltic gravel) as well as a large amount of small rocks; locally, stones appear, 20–30 cm in diameter, with individual rocks up to 50–70 cm across. Where the rocky and loamy ground is bare, it forms patches, uneven in outline and measuring from 5 × 10 up to 40 × 60 cm. The plant cover has a variegated appearance and covers *c*. 70% of the surface with white, yellowish and black surface crusts and pale-yellow *Cetraria cucullata* associated with greenish-brown mosses. The flowering plants are indistinct in the general aspect but they are actually relatively abundant (17 species; cover: 3%). This was the richest aggregation met with on the island. In comparison with the other aggregations, it had the highest number of species (see Table 7).

Nanocomplexes of type B. These consist of lichen aggregations with an important participation of species belonging to the *Cetraria cucullata* eco-group (Table 8).

The nanocomplexes belonging to this type are met with under optimal conditions of snow regime on sandy-gravelly-pebbly substrates covering the upper and middle portions of slopes on Glavnyy Val and on sandy-gravelly mineral substrates on hills in the area of Zaliv Dezhneva. Among the mosses, the abundance of which is less than in the nanocomplexes of type A, there are no species from the *Hylocomium alaskanum* eco-group. The participation of *Poa abbreviata* is characteristic; *Phippsia algida* is less

Table 8. *Floristic composition and estimated plant cover (in %) in nanocomplexes of type B[a]*

Species	25	30	33
Grasses	+	+	+
Phippsia algida	–	+	+
Poa abbreviata	+	–	–
Rushes	+	–	–
Luzula confusa	+	–	–
Herbaceous plants	2	7	4
Cardamine bellidifolia	+	–	–
Cerastium arcticum	+	+	+
Cochlearia groenlandica	+	+	+
Draba oblongata	+	+	+
D. pauciflora	+	–	–
D. subcapitata	+	+	–
Papaver polare	1	5	3
Saxifraga cernua	+	–	+
S. hyperborea	–	+	+
S. nivalis	+	–	–
Stellaria edwardsii	+	+	–
Mosses	12	12	10
Bryum arcticum	1	–	–
B. rutilans	–	2	–
Dicranoweisia crispula	–	1	–
Dicranum elongatum	1	–	–
Ditrichum flexicaule	6	+	4
Drepanocladus uncinatus	–	–	+
Encalytpa alpina	–	2	–
Hypnum revolutum	+	–	+
Pholia cruda	1	2	–
Polytrichum alpinum	2	5	4
Rhacomitrium canescens	–	–	1
Rh. lanuginosum	+	+	+
Timmia austriaca var. *arctica*	1	–	–
Tortula ruralis	–	–	+
Liverworts			
Aplozia crenulata	–	–	*
Cephaloziella arctica	*	*	*
C. hampeana	–	–	*
Lophozia alpestris	–	*	–
L. excisa	–	*	–
L. major	–	–	*
Scapania calcicola	*	*	*
S. gymnostomophila	–	*	*
S. lingulata	–	*	*
Sphenolobus minutus	*	–	–
Fruticose lichens	20	20	25
Alectoria nigricans	1	+	1
A. ochroleuca	–	1	+

Table 8 (*cont.*)

Species	25	30	33
Cetraria cucullata	–	1	8
C. ericetorum	2	7	8
C. nivalis	15	7	+
Cornicularia divergens	+	–	+
Sphaerophorus globosus	–	+	–
Stereocaulon alpinum	–	–	1
S. rivulorum	2	3	6
Tubiform lichens	+	+	–
Cladonia pyxidata (primary thallus)	–	+	–
Thamnolia vermicularis	+	–	–
Crustose lichens			
Buellia	–	–	*
Candelariella vitellina	–	*	*
Ochrolechia frigida	(*)	*	*
Lecanora campestris	(*)	(*)	*
Pertusaria sp.	–	*	*
Pyrenolichen sp.	–	*	–
Verrucaria aethibola	–	–	*
Cover of light-coloured crustose lichens	20	45	30
Cover of a black crust, thalli of crustose lichens and blue-green phycobionts as well as liverworts	5	15	20
Total plant cover	60	95	85

Note: ª For an explanation of the figures and symbols see Table 7.

frequent but there is a comparatively large number of herbs. The general plant cover amounts to 60–95%. *Papaver polare* covers 1–4% of the surface of the habitats occupied by aggregations of this type; there it not only flowers but sets fruit as well. The more satisfactory development of the poppies here than elsewhere depends not only on optimal edaphic and microclimatic conditions but also on the lesser amount of mosses (cover: *c.* 10%), among which there are many of small size (species of *Bryum*, *Pohlia*, *Encalypta* and *Timmia*) which do not form a well-developed sward that is able to suppress the growth of the poppies. Such aggregations take the form of strips that are from 1 to 10 m wide. They are easily distinguishable thanks to an abundance of bright pale-yellow cushions of *Cetraria cucullata* and *C. nivalis* and are in marked contrast to the greyish background of the habitats with a thin snow cover higher up on the slopes (Table 12) and to aggregations found lower down the slopes, where the snow persists longer and where there is much *Stereocaulon rivulorum*.

Relevé no. 25. Upper portion of a 2–3° slope with a southwest exposure near Zaliv Dezhneva. Elevation *c.* 25 m a.s.l. The snow was gone by 15

June. The substrate is a mixture of sand, gravel and small rocks. The nanorelief is slightly undulating but comparatively level. Moss–lichen strips with *Cetraria nivalis*, 5–15 cm wide, are associated with indistinctly outlined hollows. Outside the hollows mainly crustose lichens have developed. Tufts of poppies as well as *Cerastium arcticum* and small mosses grow there, too. Banks characterized by *Luzula confusa*, measuring 60–80 × 100 cm, are met with. Within the closed cover of these banks, small patches of *L. confusa* as well as mosses and *Cetraria nivalis* participate. Between them there are white and pale grey crustose lichens. Although the cover of the plants on these habitats amounts to only 60%, this is a closed cover broken by patches of bare ground only here and there.

Relevé no. 30. Upper portion of a slope on the Glavnyy Val, elevation *c.* 18 m a.s.l. The snow was gone by 15 June. The habitat stretches out like a belt and is distinguished by an abundance of brightly coloured pale yellow cushions of *Cetraria nivalis* associated with brownish cushions of *C. ericetorum* growing in rows, and with dark green tufts of poppies, which at the end of the summer of 1959 had many flowers and ripe capsules. The background is formed by pale grey and white crustose lichens together with some black crust. The general cover of plants amounts to 95%.

Relevé no. 33. Shore of Zaliv Dezhneva. Upper portion of the slope of a hill, *c.* 30 m high. The snow, the depth of which reached *c.* 10 cm, was gone by 15 June. The substrate consists of sand and gravel with a mixture of loam and a relatively large amount of stones, measuring up to 25 cm in diameter. The habitat, a longish belt the width of which is from 1 to 6 m and the length of which is more than 30 m, can be distinguished from afar thanks to the bright, pale yellow-coloured patches of *Cetraria cucullata* interspersed with brownish cushions of *C. ericetorum*. The background between them is formed by grey and black surface crusts and small patches of bare ground. The general cover of plants amounts to 85%.

Nanocomplexes of Type C. These consist of zonal moss–lichen aggregations on a loamy plakor habitat (Table 9).

A plakor habitat situated 3 km west of the Nagurskaya polar station at an elevation of *c.* 25 m a.s.l. was sampled. It has a loamy substrate, the ground is broken into polygons and is mixed with a small amount of basaltic gravel and pebbles. The relevé is situated on an elevated, worn down terrace which is *c.* 7000 years old. The loamy ground consists of ancient, clayey marine alluvium, of which the individual pebbles bear witness. The snow regime is optimal: the snow was gone by 15 June. I

Table 9. *Floristic composition and estimated plant cover (in %) within nanocomplexes of types C, D and E*[a]

Species	C	D	E
	11	23	68
Grasses	2	7	1
Phippsia algida	2	7	1
Herbaceous plants	1	2	+
Cerastium arcticum	+	–	–
C. regelii ssp. *caespitosum*	+	–	–
Draba macrocarpa	–	+	–
D. oblongata	+	–	–
D. pauciflora	+	+	–
Papaver polare	+	+	–
Saxifraga cernua	–	+	+
S. hyperborea	+	1	+
Mosses	10	20	2
Aulacomnium turgidum	+	–	–
Bartramia ithyphylla	+	–	–
Bryum rutilans	–	5	–
Campylium stellatum	+	–	–
Dicranoweisia crispula	1	–	–
Distichium capillaceum	+	–	–
Ditrichum flexicaule	5	–	–
Drepanocladus uncinatus	+	1	–
Hygrohypnum polare	+	–	+
Hylocomium alaskanum	+	–	–
Myurella apiculata	+	+	–
M. julacea	+	2	–
Orthothecium chryseum	–	5	–
O. strictum	–	1	–
Pohlia cruda	+	1	–
P. obtusifolia	–	–	+
Polytrichum alpinum	2	5	–
P. fragile	–	–	+
Psilopilum cavifolium f. *nanum*	–	–	2
Rhacomitrium canescens	+	–	–
Rh. lanuginosum	+	–	–
Tomenthypnum nitens var. *involutum*	+	–	–
Tortula ruralis	+	–	–
Liverworts			
Cephaloziella arctica	*	*	*
Gymnomitrium concinnatum	*	–	–
Lophozia alpestris	–	*	*
L. excisa	–	–	*
L. grandiretis	–	–	*
Scapania calcicola	–	–	*
S. globulifera	–	–	*
Tritomaria scitula	*	*	
Fruticose lichens	5	10	+
Cetraria cucullata	–	1	+

Table 9 (*cont.*)

Species	C	D	E
	11	23	68
C. delisei	2	1	–
C. islandica var. *polaris*	+	–	–
C. nivalis	+	+	–
Stereocaulon alpinum	+	–	–
S. rivulorum	2	9	+
Foliose lichens	+	–	–
Parmelia omphalodes	+	–	–
Tubiform lichens	+	+	+
Cladonia pocillum (horizontal thallus)	+	+	–
C. pyxidata (in relevé no. 68, squamous primary thallus)	+	–	+
Thamnolia vermicularis	+	–	–
Crustose lichens			
Caloplaca subolivacea	–	–	*
Collema ceraniscum	*	*	*
Lecanora campestris	–	–	*
Ochrolechia frigida	*	*	–
O. cf. *tartarea*	–	*	–
Pertusaria sp.	–	*	*
P. freyii	*	–	–
P. glomerata	*	–	–
P. octomela	*	–	–
Rinodina turfacea	*	–	–
Cover of light-coloured crustose lichens	45	10	+
Cover of a black crust, (see relevé 11) and black film (see relevés 23 and 68) of crustose lichens with blue-green algal phycobionts as well as liverworts	30	50	88
Total plant cover	85	99	90

Note: ^a For an explanation of the figures and symbols see Table 7.

Remarks: The species of *Pertusaria* in relevé no. 11 were identified by T. Kh. Piyn, *Collema ceraniscum* by N. S. Govorukha. The remaining lichens were identified by A. N. Oksner and M. Lamb (especially the species belonging to the genus of *Stereocaulon*).

consider this nanocomplex to be representative of a standard, zonal vegetation on this island. A polar desert surface has developed here (Fig. 28) which is broken by fissures into polygons, measuring 20–40 cm along the long axis and 15–30 cm along the short one (20 × 30 cm predominate). Greenish-brown strips of vegetation can be seen along the fissures, occupying 10–15% of the surface where mosses (e.g. *Ditrichum flexicaule*) and cushions of *Cetraria delisei* predominate and there are scattered individual specimens of flowering plants, among which are small tufts of poppies. On some of these a few flowers appear towards the end of the summer although no seeds are set. The width of the strips (Figs. 29, 30) ranges from 1 to 12 cm, is usually 3–5 cm, but is larger in places where the fissures meet. *Nostac commune* is occasionally found among the mosses in the form of small elliptic or crumb-like patches with a dark-olive colour. On the polygons patches of bare ground, occupying 10–15% of the total surface, can be found, but the major portion is covered with a crust of crustose lichens (white patches of *Pertusaria* spp. and *Ochrolechia* spp. predominate, associated with black spots of *Collema ceraniscum* and liverworts) together with small tussocks of *Phippsia algida* (mostly sterile), which are found where the fissures intersect. The general plant cover is 85%: lichens provide 80% cover and mosses 10% (the cover values resulting from the fact that some of the lichens grow on top of the mosses!). Relatively many small cushions of *Stereocaulon rivulorum* are met with both between the strips of moss–lichen sward and on the surfaces of the polygons (Fig. 30). The disconnected root systems of isolated, scattered and small individuals of flowering plants are characteristic (Fig.

Fig. 28. Polygonal polar desert on loam with predominantly crustose lichens (zonal aggregation on Zemlya Aleksandra; Relevé no. 11, Table 9) (photo.: the author).

31). In this habitat samples were collected for a study of the phytomass (Aleksandrova, 1969). The above-ground phytomass of living plants amounted to 129.1 g/m^2 (flowering plants, 6.2 g/m^2; mosses, 40.0 g/m^2; fruticose and foliose lichens, 51.7 g/m^2; crustose lichens, 30.9 g/m^2; algae, 0.3 g/m^2); the below-ground phytomass was 29.2 g/m^2. The specific composition of this aggregation is given in Table 9 (relevé no. 11).

Fig. 29. A nanocomplex at a loamy plakor habitat (Relevé no. 11, Table 9) on Zemlya Aleksandra. The snow was gone by 15 June. Size of the plot: 1.5 × 2 m. *1*, strips of moss–lichen sward along the fissures; *2*, crust of predominantly crustose lichens; *3*, bare loam.

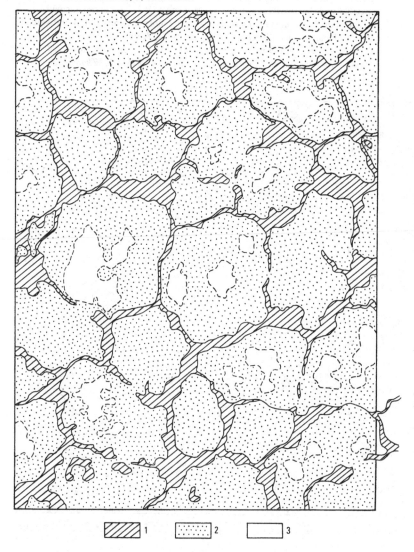

Nanocomplex of type D. This consists of tussocks of *Phippsia algida* among a black film on a moist sandy-pebbly substrate (Table 9).

This type is met with on alluvial marine terraces (elevation c. 10 m a.s.l.), has increased moisture and snow disappearing later than the optimum, e.g., by 1 July. The basic vegetation is formed by a black film consisting mainly of liverworts and crustose lichens. The presence here of miniature

Fig. 30. A 1 m² area of a nanocomplex at a loamy plakor habitat (Relevé no. 11, Table 7) on Zemlya Aleksandra. *1*, moss–lichen sward along the fissures (predominantly *Cetraria delisei* and *Ditrichum flexicaule*); *2*, crusts of mainly crustose lichens (*Pertusaria* spp., *Ochrolechia frigida*, *Collema ceraniscum*, etc.) together with liverworts (*Cephaloziella arctica*, etc.); *3*, bare loam; *4*, small cushions of *Stereocaulon rivulorum* outside the moss–lichen sward; *5*, individual shoots of *Polytrichum alpinum*; *6*, *Phippsia algida*; *7*, *Papaver polare*; *8*, *Saxifraga hyperborea*; *9*, *Cerastium arcticum*; *10*, *C. regelii* ssp. *caespitosum*; *11*, *Draba oblongata* and *D. pauciflora*; *12*, *Stellaria edwardsii*; *13*, *Thamnolia vermicularis*.

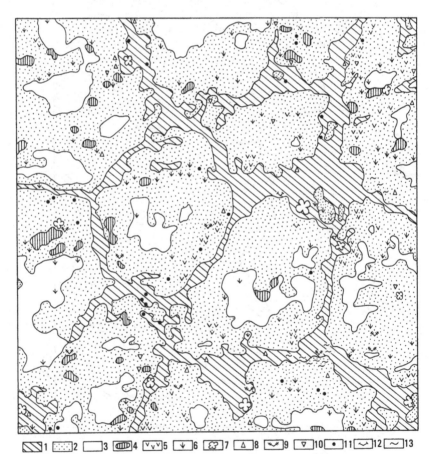

mats of *Phippsia algida*, covering up to 7% of the surface, is remarkable (concerning the tussocks of *Phippsia algida*, see below, pp. 159–61).

Relevé no. 23. A level surface at the foot of Glavnyy Val. The snow was gone by the first days of July. The substrate consists of pebbles mixed with fine soil; it was wet: the continuous vegetation (cover: 99%) is here and there disrupted by small stones. It consists almost exclusively of a velvety black film in which can be distinguished small, bright yellow spots of *Ochrolechia tartarea* and *Pertusaria* sp. Against this background can be seen strips of green, brownish and reddish mosses, forming a sward, small patches of *Stereocaulon rivulorum* and miniature, round tussocks of *Phippsia algida* (Fig. 32), 4–10 cm in diameter and 3–4 cm tall. In part they are well developed, having fertile shoots, but in part they are dead and overgrown by mosses, crustose lichens and liverworts.

Nanocomplex of type E. This consists of a black film covering wet, clayey loam (Table 9).

This type occurs occasionally on the wet, clayey-loamy deposits of ancient lagoons on low alluvial marine terraces. The dates when the

Fig. 31. Below-ground structure of a nanocomplex at a loamy plakor habitat (Relevé no. 11, Table 9) on Zemlya Aleksandra. *1*, mosses (*Ditrichum flexicaule, Polytrichum alpinum*, etc.); *2*, dead mat of *Cerastium regelii* ssp. *caespitosum* and its roots; *3*, crust formed by crustose lichens (*Pertusaria* spp., *Ochrolechia* spp., *Collema ceraniscum*, etc.); *M*, surface of permafrost at the beginning of August; *Cd, Cetraria delisei*; *Ph, Phippsia algida*; *P, Papaver polare*; *S, Saxifraga hyperborea*.

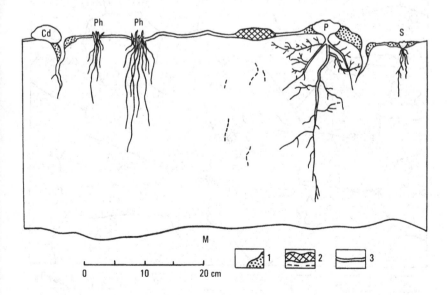

84 The vegetation

nanocomplex becomes free from snow are roughly the same as for
nanocomplex D, but the vegetation is poorer because of the clayey

Fig. 32. A nanocomplex consisting predominantly of a black film and miniature tussocks of *Phippsia algida* on Zemlya Aleksandra. The snow was gone by
the first days of July. Size of plot: 1.5 × 2 m. *1*, fissures; *2*, 'black film', locally
with addition of tiny shoots of *Psilopilum cavifolium* var. *nanum*; *3*, patches of
bare ground; *4*, tussocks of *Phippsia algida*; *5*, *Saxifraga hyperborea*; *6*, *Saxifraga cernua*; *7*, *Stereocaulon rivulorum*; *8*, *Hygrohypnum polare*.

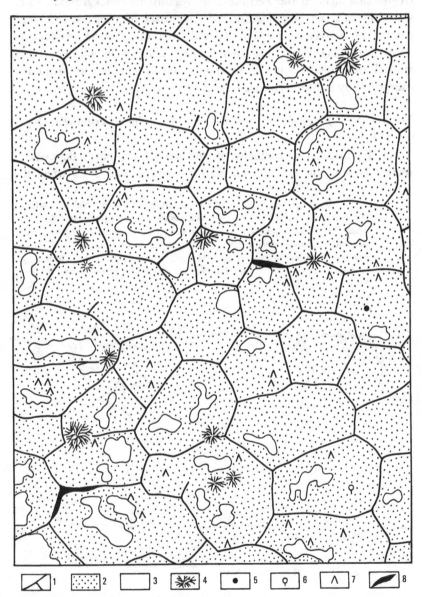

substrate and the less effective drainage in comparison to that of sandy-pebbly substrates.

Relevé no. 26. A level deposit laid down on an alluvial marine terrace in the northern part of the island. The snow disappeared some time during the first ten days of July. The ground is wet, sandy-clayey-loamy and contains a few pebbles which locally push up to the surface. Narrow fissures (from 1 mm to 1.5 cm, predominantly 0.5 cm wide) indistinctly outline the small, slightly raised polygons which are almost completely covered by a black, wrinkled film. Everywhere impoverished individual tiny shoots of *Psilopilum cavifolium* f. *nanum* can be seen; they are crowded together in groups within small depressions (cavities) of the black film and along the edges of small, loamy patches (but rarely in their centres). Along the fissures reddish *Hygrohypnum polare* is met with occasionally. Individual separate shoots of *Polytrichum fragile* are also found. Cushions of sterile *Stereocaulon rivulorum* occur in small quantities and there are tiny patches of pale reddish crustose lichens and also of small accumulations of squamules left by the initially erumpent specimens of *Cladonia pyxidata*. The general plant cover amounts to 90%. Miniature tussocks of *Phippsia algida*, 6–10 cm in diameter, appear as conspicuous components. They are barren and cover only about 1% of the surface.

Types of open aggregations (comitia)
The nanocomplexes described above always play an insignificant role in the structure of the vegetation cover on Zemlya Aleksandra. The open aggregations predominate. Those aggregations which are most widely distributed (covering over 35% of the Tsentral'naya Susha) belong to the category where individual specimens do not make contact with each other and each covers only a small part (a fraction of 1%) of the surface. The second-most widely distributed aggregations (met with in 15% of cases) are very open semi-aggregations (cover 3–15%) where, besides individuals growing alone, there are also fragments of nanophytocoenoses. About 10% of the territory is occupied by semi-aggregations with a significant plant cover. The classification, based on eco-groups in which attention is paid to the general characteristics of the plant cover, its structure and species composition, allowed me to distinguish twelve types of open aggregations. Their ecological ordination has been illustrated in Fig. 19. In the descriptions of them which follow, the most widely distributed semi-aggregations are considered first.

Open aggregations of type no. 1 (comitium no. 1). This consists of lichen aggregations on stone fields (Table 10).

Table 10.   Floristic composition and estimated plant cover (in %) in open aggregations belonging to type no. 1 (Comitium no. 1, subcomitia nos. 1' and 1'')ᵃ

Species	1'				1''	
	5	6	16	21	29	70
Grasses						
Phippsia algida	—	—	—	+	—	—
Herbaceous plants						
Cerastium arcticum	—	—	—	+	—	—
C. regelii ssp. caespitosum	—	—	—	+	+	+
Cochlearia groenlandica	—	—	—	—	—	—
Draba macrocarpa	—	—	—	+	—	+
D. oblongata	—	—	—	+	—	+
D. subcapitata	—	—	—	+	—	+
Papaver polare	—	—	—	—	+	+
Saxifraga cernua	—	—	—	+	+	—
S. hyperborea	—	—	—	—	+	—
Mosses						
Andreaea rupestris	+	1	2	13	10	1
Bartramia ithyphylla	—	—	+	1	1	—
Bryoerythrophyllum ferruginascens	—	—	—	+	—	—
Dicranoweisia crispula	—	+	+	+	2	—
Dicranum elongatum	+	+	—	10	—	—
Ditrichum flexicaule	—	—	+	—	+	+
Myurella julacea	—	+	—	—	—	—
Orthothecium strictum	—	—	+	+	—	—
Pohlia cruda	—	+	—	—	—	—
Polytrichum alpinum	—	+	—	+	—	—

Species	1	2	3	4	5	6
Rhacomitrium lanuginosum	+	–	+	1	1	+
Timmia austriaca var. *arctica*	–	–	–	+	–	–
Liverworts						
Anthelia juratzkana	*	*	*	*	*	*
Cephaloziella arctica	*	*	*	*	*	*
Gymnomitrium concinnatum	–	–	–	–	–	–
G. corallioides	*	*	*	*	*	*
G. obtusum	*	*	*	*	*	*
Lophozia excisa	–	–	–	–	–	–
Scapania calcicola	–	–	–	–	–	–
Sphenolobus minutus	*	*	*	*	*	*
Tritomaria scitula	*	*	*	*	*	*
Fruticose lichens						
Alectoria jubata var. *chalybeiformis*	15	20	10	9	20	5
A. minuscula	–	–	–	–	–	–
A. nigricans	+	+	+	+	+	+
A. nitidula	–	–	–	–	–	–
A. ochroleuca	+	+	+	+	+	+
A. pubescens	1	–	1	–	2	–
Cetraria cucullata	+	+	+	–	1	–
C. delisei	1	+	–	+	–	–
C. ericetorum	+	–	+	+	–	+
C. hepatizon	–	+	+	+	–	–
C. islandica var. *polaris*	+	–	–	–	–	–
C. nigricans	5	–	–	–	–	–
C. nivalis	+	–	1	–	–	–
Cornicularia aculeata	1	–	–	+	–	–
C. divergens	+	–	–	+	+	+
Sphaerophorus fragilis	1	+	+	+	+	+
S. globosus						+

Table 10 (cont.)

		1′			1″	
Species	5	6	16	21	29	70
Stereocaulon alpinum	–	–	–	–	+	+
S. botryosum f. *congestum*	–	–	–	3	–	–
S. rivulorum	–	–	+	3	–	–
S. vesuvianum var. *pseudofastigiatum*	3	4	+	–	2	3
Usnea sulphurea	2	13	7	+	11	1
Foliose lichens						
Hypogymnia intestiformis	25	15	10	6	7	12
H. physodes	+	+	+	–	–	–
Parmelia omphalodes	1	1	–	–	–	–
Peltigera rufescens var. *incusa*	–	–	+	+	–	–
Umbilicaria arctica	–	–	–	6	–	2
U. proboscidea	25	14	10	6	7	10
Tubiform lichens						
Cladonia pocillum (horizontal thallus)	+	+	+	+	+	+
C. pyxidata	–	–	–	+	–	–
Thamnolia vermicularis	+	+	+	+	+	+
Crustose lichens growing on soil						
Buellia	5	10	10	20	25	50
Caloplaca elegans	–	–	–	–	–	*
Lecanora campestris	–	–	–	–	–	*
L. polytropa	*	*	–	–	–	*
Lecidea dicksonii	*	*	*	*	*	*
L. macrocarpa	–	*	–	*	*	*
Rhizocarpon geographicum	*	*	–	–	–	–

Verrucaria aethiobola	*	—	*	—	—	—
Crustose lichens growing on soil and remnants of plants						
Candellariella vitellina	—	—	—	—	—	—
Ochrolechia frigida	*	*	*	*	(*)	(*)
O. cf. tartarea	*	*	*	*	—	—
Pannaria pezizoides	*	—	—	—	—	—
Pertusaria sp.	*	—	*	*	—	—
Rinodina turfacea	*	*	—	—	—	—
Cover of light-coloured crustose lichens growing on soil and remnants of plants	25	10	6	20	35	45
Cover of a black crust of crustose lichens, blue-green algal phycobionts and liverworts, growing on remnants of plants	25	10	6	3	5	5
Total plant cover	85	60	45	70	80	80

Note: [a] For an explanation of the figures and symbols see Table 7.

This type of open aggregation comprises semi-aggregations developed on habitats with piles of rocks and a thin or moderate snow cover. The accumulations of stones on which they are found consist in the majority of cases of angular blocks of basalt and only occasionally of boulders of morainic origin (relevé no. 5). Although the plant cover in some cases amounts to 85%, there is no closed cover here, i.e., the stones overgrown with lichens are isolated from each other by intervals between them. Two subtypes can be distinguished: 1', where flowering plants are entirely absent, and 1", where individual specimens of flowering plants are met with.

Subtype 1' (subcomitium 1'). This consists of lichen aggregations on stone fields without any flowering plants.

Relevé no. 5. The top and upper portion of a slope on a low hill, 5 km south of the Nagurskaya polar station. The snow was gone by 10 June although the mean daily temperature still held steady below 0°C. Stones with a diameter from 5–7 up to 30 cm are densely covered with lichens among which the yellow thalli of *Umbilicaria proboscidea* predominate. Mosses are few and grow together with some fruticose lichens in the spaces between the stones. There are no flowering plants.

Relevé no. 6. A conglomeration of stones forming a network at a plakor habitat, elevation *c*. 25 m a.s.l. The snow was gone by 12 June. The width of the stone belts is 0.5–2.5 m. Chaotic accumulations of basaltic rocks measuring mainly 15 × 25 cm (but up to 50 × 80 cm) occur; occasionally individual boulders reaching 1 m in width and 50 cm in height can be seen. Among the plants, which cover about 60% of the stone surface, black *Umbilicaria* and greenish tufts of *Usnea sulphurea* predominate. The presence of black, finely tufted *Alectoria pubescens* is also characteristic. Mosses are few and hide between the stones.

Relevé no. 16. The upper portion of a very gentle slope with a southern exposure at *c*. 25 m elevation a.s.l. The snow was gone by 12 June. Disorderly accumulations of stones form a network. Blocks of basalt measure 40 × 50 cm, 20 × 35 cm or less, individual boulders reach 80 × 120 cm. On the surface of the stones black thalli of *Umbilicaria*, green tufts of *Usnea sulphurea*, small pale patches of crustose lichens and black cushions of *Alectoria pubescens* can be seen; black mats of *Andreaea rupestris* occur among the mosses on the surface of the stones. A few other species of mosses are crowded in between the stones. There are no flowering plants.

Subtype 1" (subcomitium 1"). This consists of lichen aggregations on stones with the occasional participation of flowering plants.

Relevé no. 21. The upper portion of a gentle slope at *c.* 25 m elevation a.s.l. The snow was gone by 20 June. The substrate is an accumulation of basaltic blocks measuring 15 × 20 cm and 20 × 25 cm with many small stones in the spaces between them. There are black patches of *Umbilicaria*, compact black mats of *Andreaea*, small white spots of crustose lichens (a part of which are *Ochrolechia* and *Pertusaria* forming a crust on the remnants of dead specimens of *Andreaea*), *Alectoria pubescens*, etc., on the surfaces of the stones. The mosses are more abundant here than in aggregations of subtype 1′, cushions of *Discranoweisia* predominating. There are no flowering plants.

Relevé no. 29. A stone net (see Fig. 27) on a level habitat at *c.* 25 m elevation a.s.l. The snow was gone by 20 June. The diameter of the stones measures 10–35 cm; individual boulders reach 60–80 cm across. Among the lichens the characteristic ones are basically the same as are listed for relevé no. 21. *Rhacomitrium lanuginosum* is the commonest moss. Flowering plants are found but in very small amounts. The general cover is 80%.

Relevé no. 70. A site along Zaliv Dezhneva at *c.* 35 m elevation a.s.l. The substrate consists of an eluvium deposited on basaltic boulders; locally the base rock itself outcrops. The snow was gone by 20 June. The plants cover about 80% of the surfaces of the stones; epilithic crustose lichens predominate and there is relatively much of *Umbilicaria*. The mosses are very sparse: mainly *Rhacomitrium lanuginosum* is met with. There are only individual flowering plants.

Open aggregations of type no. 2 (comitium no. 2). This consists of lichen–moss aggregations with *Rhacomitrium lanuginosum* on stone fields (Table 11).

This type is characterized by aggregations (*comitia*), developed on heaps of stones, with a greater abundance of *Rhacomitrium lanuginosum* than in aggregations belonging to type no. 1, although with respect to the eco-groups their structure is similar.

Relevé no. 14. A gentle slope facing north-east. Elevation *c.* 18 m a.s.l. The snow was gone by 22 June. The stone net is composed of basaltic blocks up to 70 cm in diameter (predominantly 24–40 cm). Uneven rows and strips of *Rhacomitrium lanuginosum* are found on the stone net, covering about 40% of the surface. The general plant cover amounts to 60%. Green cushions of *Oncophorus wahlenbergii* can also be seen. There are a few lichens but only individual flowering plants.

Open aggregations of type no. 3 (comitium no. 3). This consists of crusts of lichens on sandy-pebbly and sandy-gravelly habitats with some fine soil and a scanty snow cover (Table 12).

Table 11. *Floristic composition and estimated plant cover (in %) in open aggregations of type no. 2 (Comitium no. 2)*[a]

Species	14
Grasses	+
Phippsia algida	+
Herbaceous plants	+
Cerastium regelii ssp. *caespitosum*	+
Draba pauciflora	+
Mosses	45
Andreaea rupestris	+
Ditrichum flexicaule	+
Drepanocladus uncinatus	+
Oncophorus wahlenbergii	4
Rhacomitrium lanuginosum	40
Liverworts	
Blepharostoma trichophyllum	*
Cephaloziella arctica	*
Sphenolobus minutus	*
Fruticose lichens	5
Alectoria minuscula	+
A. nigricans	1
A. pubescens	+
Cetraria cucullata	+
C. ericetorum	1
C. islandica var. *polaris*	+
Cornicularia divergens	1
Stereocaulon alpinum	+
Usnea sulphurea	1
Foliose lichens	1
Umbilicaria proboscidea	1
Tubiform lichens	+
Cladonia pyxidata	+
Crustose lichens, growing on leaves	5
Lecidea dicksonii	*
L. macrocarpa	*
Rhizocarpon geographicum	*
Crustose lichens growing on soil and remnants of plants	
Ochrolechia frigida	*
Pertusaria sp.	*
Rinodina turfacea	*
Cover of light-coloured crustose lichens	3
Cover of a black crust	2
Total plant cover	60

Note: [a] For an explanation of the figures and symbols see Table 7.

Within this type are included semi-aggregations which have developed on sites where the snow disappears early (at the end of May or the beginning of June) and where the ground consists of sandy-pebbly or sandy-gravelly-fine mineral soils. In these aggregations species mainly from the *Stereocaulon vesuvianum* var. *depressum* eco-group participate. Although the winter conditions are severe here, the development of the plants during summer is favoured by a longer growing period compared with that at other ecotopes and by a higher sum total of positive temperatures. Consequently – but also due to the presence of 'stone greenhouses', where the favourable position of stones protects the individual specimens from wind and provides them with more heat during the daylight hours than does the fine soil (see p. 24) – in such habitats there are relatively more species, which not only flower but also produce ripe seeds. These habitats act as a special kind of 'seed dispersal centre', supplying seeds spread by the winds to sites where the plants are unable to produce ripe seeds. In these habitats *Papaver polare* in particular flowers and produces ripe seeds just as it does on sites belonging to the nanocomplex of type B. Because of the scanty snow cover and the relatively dry conditions, these localities are not favourable for the existence of *Phippsia algida;* however, *Poa abbreviata* thrives there and produces ripe seeds. Mosses are only weakly developed, being represented in general by such small plants as *Encalypta alpina*, species of *Bryum* and *Pohlia* and also by small, individual cushions of *Dicranum elongatum*. Among the lichens the crustose forms predominate, forming surface crusts in which *Stereocaulon vesuvianum* var. *depressum* frequently plays a prominent part. There are few other lichens. Although the cover is in general high, there is no closed vegetation cover here, since when the crusts over the soil are disrupted they do not grow together again.

Relevé no. 20. Top of ridge, *c.* 30 m a.s.l., at Zaliv Dezhneva. The snow was gone by the end of May. A level surface built up of sand and gravel with a mixture of loamy patches. Spots and strips of a grey crust of lichens and individual specimens of flowering plants (among them, poppies with a cover of 2%) are scattered over the bare ground. The crustose lichens are frequently mixed with small mosses. The general plant cover amounts to 30%.

Relevé no. 31. The top (crest) of Glavnyy Val. The snow was gone by early June. The ground consists of a mixture of sand and pebbles, mainly of a small size; individual pebbles are found measuring up to 20 cm across. There are many crusts of grey, crustose lichens, and *Stereocaulon vesuvianum* var. *depressum* is relatively abundant here. Individual tufts of herbs can be seen, mostly of *Papaver polare*. The general plant cover amounts to 65% (of which 60% consists of surface crusts of crustose lichens).

Table 12. Floristic composition and estimated plant cover (in %) in open aggregations belonging to type no. 3 (Comitium no. 3)[a]

Species	20	31	32	34	36
Grasses					
Phippsia algida	+	+	+	–	+
Poa abbreviata	–	+	–	–	–
Herbs					
Cerastium arcticum	+	+	+	+	+
Cochlearia groenlandica	5	+	+	+	+
Draba oblongata	1	+	+	+	+
D. pauciflora	1	+	–	+	+
D. subcapitata	+	+	+	–	+
Minuartia rubella	+	–	–	–	–
Papaver polare	+	–	+	–	+
Saxifraga caespitosa	+	+	+	+	–
S. cernua	2	–	+	–	+
S. hyperborea	1	–	+	–	+
S. nivalis	+	+	+	+	+
Stellaria edwardsii	+	+	–	+	–
Mosses					
Bartramia ithyphylla	1	1	1	1	5
Bryum pseudocrispulum	–	+	–	–	–
B. rutilans	–	–	–	–	+
Dicranoweisia crispula	+	+	–	+	2
Dicranum elongatum	–	–	–	+	–
Ditrichum flexicaule	+	–	–	+	1
Drepanocladus uncinatus	–	–	–	+	–
Encalypta alpina	+	+	+	+	–

Species					
Hypnum revolutum var. pygmaeum	+	—	+	—	—
Kiaeria starkei	—	+	—	+	—
Orthothecium strictum	—	+	—	—	—
Pohlia cruda	—	—	+	—	—
Polytrichum alpinum	+	—	—	+	—
Rhacomitrium canescens	+	—	—	—	—
Rh. lanuginosum	—	+	+	—	—
Timmia austriaca var. arctica	—	—	—	—	+
Tortula ruralis	—	+	+	—	—
Liverworts					
Aplozia crenulata	•	—	—	—	—
Cephaloziella arctica	•	—	•	•	•
C. hampeana	—	•	—	—	—
Jungermannia polaris	—	•	—	—	—
Lophozia excisa	•	—	—	—	—
L. longidens var. arctica	—	—	—	•	—
Scapania calcicola	•	—	—	•	—
S. globulifera	—	•	—	—	—
S. mucronata	—	•	—	—	—
S. praetervisa	—	—	—	•	—
Sphenolobus minutus	•	—	—	—	•
Tritomaria scitula	—	•	—	—	•
Fruticose lichens					
Alectoria nigricans	20	+	+	5	+
A. ochroleuca	—	—	+	—	+
A. pubescens	—	+	—	—	—
Cetraria cucullata	—	—	+	—	—
C. ericetorum	+	+	+	—	+
C. nivalis	+	+	+	—	+
Stereocaulon alpinum	+	—	—	1	—
S. rivulorum	—	+	—	—	—

Table 12 (cont.)

Species	20	31	32	34	36
S. vesuvianum var. *depressum*	–	4	+	–	20
Foliose lichens					
Parmelia omphalodes	+	–	+	–	–
Umbilicaria proboscidea	+	–	+	–	–
Tubiform lichens					
Thamnolia vermicularis	+	–	+	+	+
Crustose lichens					
Buellia sp.	–	*	*	–	–
Caloplaca jungermanniae	–	*	–	–	–
C. stillistidiorum	–	–	*	*	*
Collema ceraniscum	*	–	–	–	–
Lecanora campestris	*	–	*	*	*
Lecidea sp.	*	–	–	–	–
Ochrolechia frigida	*	*	(*)	*	*
Ochrolechia sp.	–	*	*	*	*
Pertusaria sp.	*	–	–	*	*
Verrucaria aethiobola	*	–	–	–	–
Cover of light-coloured crustose lichens	20	55	20	70	50
Cover of a black film of crustose lichens, blue-green phycobionts and liverworts	5	5	5	10	10
Total plant cover	30	65	27	90	85

Notes: [a] For an explanation of the figures and symbols see Table 7.

Relevé no. 32. Northern slope and top of a hill, elevation *c*. 35 m a.s.l., at Zaliv Dezhneva. The snow was gone by the end of May. Gravel mixed with sand and a small amount of stone measuring 5–12 cm predominates. Where the base rock outcrops in the form of basalt blocks, 0.5–1.5 m wide, black thalli of *Umbilicaria* grow. On the fine soil uneven patches of pale grey crustose lichens and brownish *Stereocaulon vesuvianum* var. *depressum* with small inclusions of black crust can be seen. Various small patches of mosses are present and, separately, mats of herbs as well as of *Poa abbreviata* are met with. The general cover is *c*. 30%.

Relevé no. 34. Small area exposed to winds on a west-facing slope near Zaliv Dezhneva. Elevation *c*. 25 m a.s.l. The snow was gone by the beginning of June. The substrate consists of sand mixed with basaltic gravel and a minor amount of small stones. The surface is both level and rolling; it has a large amount of crust formed by pale grey crustose lichens (e.g. *Lecanora campestris*) covering 80% of the surface. There are very few species of moss or flowering plants.

Relevé no. 36. Same area; a 2–3° slope with a southerly exposure. The snow was gone by 10 June. The substrate consists of sand and gravel mixed with small rocks. About 50% of the surface is covered by crusts of pale grey crustose lichens among which grey *Stereocaulon vesuvianum* var. *depressum* is prominent. There are also patches and strips of black crust as well as of white crustose lichens and individual mats of poppies and small mosses. Separate specimens of *Poa abbreviata* are also met with.

Open aggregations of type no. 4 (comitium no. 4). This consists of strips and mats of a moss (*Ditrichum flexicaule*, etc. – lichen semi-aggregation on loam broken into polygons (Table 13).

In this *comitium* we can observe the same kind of polygonal fissuring of the ground as is found with the nanocomplex of Type B. However, the snow disappeared about a week later here and so there is less heat. The vegetation consequently does not form a closed cover but is represented by a semi-aggregation consisting of interrupted strips of mosses along the fissures and fragments of crust on the soil. Where the ground is wetter, there is much *Phippsia algida* (cover: 7%) in the form of very small, mostly sterile tussocks.

Relevé no. 8. A gentle (1–2°) slope at an elevation of *c*. 25 m a.s.l., south-exposed and bordering on a plakor habitat, *c*. 3 km west of the Nagurskaya polar station. The snow was gone by 22 June. The substrate is loamy with a slight mixture of gravel and small stones and is broken into polygons, measuring 20 × 25 to 25 × 30 cm. Along the fissures there are discontinuous strips of mosses with a few patches of flowering plants.

Table 13. *Floristic composition and estimated plant cover (in %) in open aggregations belonging to type no. 4 (Comitium no. 4)[a]*

Species	8	12
Grasses	1	7
Phippsia algida	1	7
Herbs	3	+
Cerastium arcticum	1	−
C. regelii ssp. *caespitosum*	+	+
Cochlearia groenlandica	+	−
Draba oblongata	+	+
D. pauciflora	+	+
D. subcapitata	+	−
Papaver polare	1	+
Saxifraga cernua	+	−
S. hyperborea	−	+
Stellaria edwardsii	+	+
Mosses	15	10
Aulacomnium turgidum	1	−
Bryum rutilans	1	−
Campylium stellatum	−	+
Cirriphyllum cirrosum	+	−
Dicranoweisia crispula	−	1
Ditrichum flexicaule	8	4
Drepanocladus uncinatus	1	+
Hygrohypnum polare	−	+
Hypnum revolutum var. *pygmaeum*	1	−
Polytrichum alpinum	2	4
P. fragile	−	+
Rhacomitrium canescens	−	+
Liverworts		
Aplozia crenulata	*	−
Cephaloziella arctica	*	*
Fruticose lichens	7	10
Cetraria cucullata	1	−
C. delisei	−	2
C. ericetorum	1	−
Stereocaulon rivulorum	5	8
Foliose lichens	+	−
Solorina bispora	+	−
Tubiform lichens	+	+
Cladonia pyxidata	+	−
Thamnolia vermicularis	−	+
Crustose lichens		
Caloplaca jungermanniae	*	−
Collema ceraniscum	*	(*)
Lecidea sp.	*	−
Ochrolechia frigida	*	*
Pertusaria sp.	*	*
Rinodina turfacea	*	(*)

Table 13 (*cont.*)

Species	8	12
Cover of light-coloured crustose lichens	17	13
Cover of a black crust of crustose lichens, blue-green algae and liverworts	8	12
Total plant cover	50	35

Note: *ª* For an explanation of the figures and symbols see Table 7.

About 50% of the surface of the polygons is covered by a black-and-white crust of crustose lichens, very small tufts of *Stereocaulon rivulorum* and various kinds of flowering plants.

Relevé no. 12. Same area, but a wetter site. The snow was gone by 20 June. The substrate is broken by narrow fissures into polygons 15–35 cm broad and 25–40 cm long. The cracks are not overgrown by vegetation. On the bare loam black and white crusts appear together with greenish-brown strips and patches of mosses or of *Cetraria delisei*, a few tufts of *Stereocaulon rivulorum* and many remnants of sterile *Phippsia algida* tussocks. The general cover amounts to 35%.

Open aggregations of type no. 5 (comitium no. 5). This consists of a semi-aggregation of moss–lichen sward with *Cetraria delisei* and fragments of a black film on wet, stony ground mixed with fine soil (Table 14).

The large number of stones due to cryogenic sorting is characteristic, as are the increased amount of moisture, a significant quantity of *Cetraria delisei*, a good cover by fragments of black film and, in a few cases, a relatively large abundance of *Phippsia algida*.

Relevé no. 19. A habitat situated not far from the shore of a lake at an elevation of about 20 m a.s.l. The snow was gone by 20 June. The ground consists of accumulations of basaltic blocks, measuring 10–15 (up to 25 × 40) cm, among which are interspersed small medallions (from 30 × 44 to 40 × 70 cm) of smaller rocks with a mixture of loam. On these fragments of black film a little moss, *Stereocaulon rivulorum* and individual tussocks of *Phippsia algida* are found. Here and there occurs a single specimen of *Papaver polare* or some other flowering plants. Cushions of *Rhacomitrium lanuginosum, Dicranoweisia crispula* and *Cetraria delisei* are scattered among the stones. The general cover is 60%.

Relevé no. 24. A habitat with an almost horizontal surface at about 20 m elevation a.s.l. The snow was gone by 22 June. The ground is covered by stones measuring 15 × 20 up to 60 × 80 cm and smaller rocks as well as some addition of loam. The jumbled appearance is due to sorting. Where there are larger stones cushions of *Dicranoweisia crispula* and

Table 14. *Floristic composition and estimated plant cover (in %) in open aggregations belonging to type no. 5 (Comitium no. 5)[a]*

Species	19	24	26	46
Grasses	+	5	+	1
Phippsia algida	+	5	+	1
Herbs	+	+	+	1
Cerastium regelii ssp. caespitosum	–	+	–	+
Cochlearia groenlandica	+	–	+	+
Draba oblongata	+	–	–	+
D. pauciflora	–	+	–	–
D. subcapitata	–	–	–	+
Papaver polare	+	+	–	+
Saxifraga cernua	–	–	–	+
S. hyperborea	+	+	+	+
S. oppositifolia	–	+	–	–
Stellaria edwardsii	+	–	+	–
Mosses	15	15	7	15
Bartramia ithyphylla	+	–	–	–
Bryum pseudocrispulum	–	–	–	+
B. rutilans	+	+	+	+
Campylium stellatum	–	+	+	–
Dicranoweisia crispula	3	6	5	8
Distichium capillaceum	–	+	–	+
Ditrichum flexicaule	–	+	–	4
Drepanocladus revolvens	+	–	–	–
D. uncinatus	–	+	–	–
Hygrohypnum polare	–	+	–	–
Kiaeria starkei	–	–	+	–
Mnium rugicum	–	–	+	–
Oncophorus wahlenbergii	+	–	1	2
Orthothecium chryseum	+	+	+	+
Pohlia cruda	–	–	–	+
Polytrichum alpinum	2	–	–	–
P. fragile	–	8	–	–
Rhacomitrium canescens	–	+	–	–
Rh. lanuginosum	10	–	–	–
Timmia austriaca var. arctica	–	–	–	+
Liverworts				
Aplozia crenulata	–	–	–	*
Blepharostoma trichophyllum	–	–	–	*
Cephaloziella arctica	*	*	*	*
Lophozia excisa	*	–	–	–
Orthocaulis kunzeanus	–	–	–	*
O. quadrilobus	–	–	–	*
Tritomaria scitula	–	–	–	*
Fruticose lichens	5	10	10	15
Alectoria nigricans	–	–	–	+
A. ochroleuca	–	–	–	+
Cetraria cucullata	–	+	–	+

Table 14 (*cont.*)

Species	19	24	26	46
C. delisei	4	13	4	7
C. ericetorum	–	–	–	1
C. nivalis	+	–	–	+
Cornicularia divergens	+	–	–	+
Stereocaulon rivulorum	1	2	6	6
Usnea sulphurea	–	+	–	–
Foliose lichens	–	+	–	+
Solorina bispora	–	–	–	+
Umbilicaria proboscidea	–	+	–	–
Tubiform lichens	–	+	–	+
Cladonia pocillum (horizontal thallus)	–	–	–	+
Cladonia pyxidata (squamous primary thallus)	–	–	–	+
Thamnolia vermicularis	–	+	–	+
Crustose lichens				
Caloplaca jungermanniae	*	–	–	–
C. stillistidiorum	–	–	–	*
C. subolivacea	*	–	–	–
Candelariella vitellina	–	–	–	*
Collema ceraniscum	*	(*)	(*)	*
Lecidea sp.	–	–	–	*
Ochrolechia frigida	*	(*)	(*)	*
O. cf. *tartarea*	–	–	–	*
Pertusaria sp.	(*)	(*)	(*)	*
Rinodina turfacea	*	(*)	–	–
Cover of light-coloured crustose lichens	5	15	3	5
Cover of a black crust of crustose lichens, blue-green algae and liverworts	25	15	20	25
Total plant cover	60	50	30	60

Note: ª For an explanation of the figures and symbols see Table 7.

Cetraria delisei are found, and on the fine soil there are crusts of crustose lichens, fragments of black film, an accumulation of *Polytrichum fragile* shoots and other small mosses as well as small tussocks of *Phippsia algida* covering *c.* 5% of the surface and single representatives of other herbs (Fig. 33). The general plant cover amounts to 50%.

Relevé no. 26. A gently sloping lake shore at *c.* 10 m elevation a.s.l. near Zaliv Dezhneva. It is occupied by a wet network of stones, 7 × 10 to 25 × 30 cm in size. The snow disappeared during the second half of June. *Cetraria delisei* and tufts of *Dicranoweisia crispula* grow on the stones, and along the edges of the medallions of fine soil there are crustose lichens and fragments of black film as well as some moisture-loving mosses; the centre of the medallions is bare with individual patches of crust on the soil, tussocks of *Phippsia algida* and other herbs. The general cover amounts to 30%.

Relevé no. 46. A gentle slope at *c.* 20 m elevation a.s.l. towards the northeast at Zaliv Dezhneva. The ground consists of small rocks the size of which reaches 15 × 20 cm. Rough patches of bare ground alternate with black and white crust on the soil, strips and patches of *Cetraria delisei* and mosses; here and there small tussocks of *Phippsia algida* and small herbs can be seen. The general cover is 60%.

Open aggregations of type no. 6 (comitium no. 6). This consists of semi-aggregations with cushions and patches of *Stereocaulon rivulorum* on a gravelly or gravelly-mineral soil (Table 15).

The semi-aggregations belonging to this type are distributed on sites where the snow disappears between 1 and 15 July, i.e., later than the optimum time. They occur on roughly 3% of the area of the Tsentral'naya

Fig. 33. Semi-aggregation with a predominance of cushions of *Cetraria delisei*, *Dicranoweisia crispula* and a crust of crustose lichens on Zemlya Aleksandra; on the stones various thalli of *Umbilicaria proboscidea* and fruticose *Usnea sulphurea* can be seen (Relevé no. 24, Table 14) (photo.: the author).

Table 15. Floristic composition and estimated plant cover (in %) in open aggregations belonging to type no. 6 (Comitium no. 6)[a]

Species	22	37	38	39	40	47	55	63	64	65	69
Grasses											
Phippsia algida	2	+	+	+	+	1	3	1	1	+	+
	2	+	+	+	+	1	3	1	1	+	+
Herbs											
Cardamine bellidifolia	–	+	1	1	1	1	1	3	3	1	1
Cerastium arcticum	+	+	–	–	–	–	–	–	–	+	–
C. regelii ssp. caespitosum	–	+	+	+	+	+	–	–	1	+	+
Cochlearia groenlandica	–	+	1	–	–	–	+	1	–	+	+
Draba macrocarpa	–	–	+	+	+	+	+	+	+	–	–
D. oblongata	+	+	–	–	–	+	–	+	+	+	+
D. pauciflora	–	+	+	+	+	–	+	–	–	–	–
D. supcapitata	+	+	+	+	+	+	+	+	+	+	+
Papaver polare	+	+	+	+	+	+	+	+	+	+	+
Saxifraga cernua	–	+	+	+	+	+	+	1	1	+	+
S. hyperborea	–	–	+	+	+	+	+	1	1	+	–
Stellaria edwardsii	+	–	–	+	–	+	–	+	+	+	–
Mosses											
Bryum pallescens	10	5	10	1	3	5	15	12	10	5	10
B. pseudocrispulum	–	+	–	–	–	–	–	–	–	–	–
B. rutilans	1	1	–	–	1	–	–	+	–	+	+
B. teres	–	–	–	+	–	+	–	+	–	+	1
B. tortifolium	–	–	4	–	–	–	–	–	–	–	–
Dicranoweisia crispula	–	–	–	–	–	+	2	+	+	+	–
Dicranum elongatum	–	–	–	–	+	–	–	–	+	–	–
Ditrichum flexicaule	–	–	–	–	–	–	–	–	–	–	–
Drepanocladus uncinatus	+	+	–	+	–	–	–	–	–	–	+
Hygrohypnum polare	–	–	–	–	–	+	–	–	–	–	–

Table 15 (cont.)

Species	22	37	38	39	40	47	55	63	64	65	69
Kiaeria starkei	—	—	—	—	1	—	—	—	—	—	—
Myurella julacea	—	—	—	—	—	—	—	+	—	—	+
Oncophorus wahlenbergii	—	—	—	—	—	—	—	—	—	—	—
Orthothecium chryseum	—	—	—	—	—	—	+	—	—	—	—
Pogonatum urnigerum var. *subintegrifolium*	1	1	—	—	—	—	—	—	+	—	1
Pohlia cruda	—	—	—	—	—	—	2	3	2	1	3
Pohlia nutans	—	—	—	+	+	—	—	—	—	—	—
Polytrichum alpestre	+	—	—	+	+	—	—	—	—	—	3
P. alpinum	7	2	6	—	—	4	10	8	7	1	—
P. fragile	—	—	—	—	—	—	—	—	—	2	—
Psilopilum cavifolium f. *nanum*	—	+	—	—	—	+	—	—	+	—	—
Rhacomitrium canescens	—	—	—	+	—	—	+	+	+	—	+
Tortula ruralis	—	—	—	—	—	—	+	—	—	—	—
Liverworts											
Aplozia cremulata	*	*	—	*	*	*	*	*	*	—	*
Cephaloziella arctica	—	*	*	—	*	—	—	—	—	*	*
Lophozia alpestris	—	—	—	—	*	—	—	—	—	*	*
L. excisa	—	—	—	*	*	*	—	—	*	—	*
L. grandiretis	—	*	—	—	—	*	—	—	—	—	—
Scapania calcicola	—	*	—	—	—	*	*	*	*	—	—
S. globulifera	—	—	—	—	—	—	—	—	—	—	—
Sphenolobus minutus	—	—	—	—	*	—	—	—	—	—	—
Fruticose lichens	45	40	10	50	50	25	30	25	12	35	20
Alectoria nigricans	—	—	—	—	—	+	+	+	—	—	—
Cetraria cucullata	—	—	—	—	—	+	+	+	—	—	—
C. delisei	+	+	+	—	—	—	—	—	—	—	—
C. ericetorum	—	—	—	—	—	—	—	—	—	—	—

C. islandica var. polaris	–	–	–	–	–	–	–	–	–	–	+
C. nivalis	+	+	+	+	+	+	+	+	–	–	–
Cornicularia divergens	–	–	–	–	+	+	–	–	–	–	–
Stereocaulon alpinum	–	–	–	–	–	–	–	–	–	–	+
S. rivulorum	45	40	10	50	50	24	30	25	12	35	20
Foliose lichens											
Peltigera erumpens	+	–	–	–	–	–	+	+	–	–	+
Solorina bispora	+	–	–	–	–	–	+	–	–	–	+
Tubiform lichens											
Cladonia pocillum (horizontal thallus)	–	–	–	–	+	+	+	+	–	+	–
C. pyxidata (in relevé 65, squamous primary thallus)	–	–	–	–	+	+	+	–	–	–	–
Thamnolia vermicularis	–	–	–	–	–	–	+	+	–	–	–
Crustose lichens											
Candellariella vitellina	*	–	(*)	–	*	*	*	–	–	–	–
Collema ceraniscum	–	–	–	–	*	*	(*)	*	–	–	*
Lecidea sp.	(*)	–	–	–	–	*	–	–	–	–	–
Ochrolechia cf. tartarea	*	–	*	–	(*)	*	*	*	–	*	–
Ochrolechia sp.	(*)	*	–	–	*	*	–	(*)	–	*	–
Pertusaria sp.	(*)	–	–	–	*	*	(*)	(*)	*	–	*
Cover of light-coloured crustose lichens	10	2	2	+	5	5	5	8	4	7	5
Cover of a black crust of crustose lichens, blue-green phycobionts and liverworts as well as fragments of 'black film' consisting mainly of liverworts	15	18	28	5	15	20	25	22	7	3	15
Total plant cover	80	60	50	55	75	55	80	80	35	45	50

Note: [a] For an explanation of the figures and symbols see Table 7.

Susha. Ash-grey patches of *Stereocaulon rivulorum* are the most characteristic component, sprinkled with shades of brown due to the brown colour
of the apothecia. The size ranges from small (but many) cushions up to
small mats, the width of which in some cases amounts to 50–100 cm and
the length of which can be from 60 up to 200 cm. They are separated from
each other by bare ground. Mosses play an unimportant role and occur
mostly as spots of black crust. Finally, there are some specimens of
Phippsia algida, but these are few in number.

Relevé no. 22. A site at an elevation of *c*. 12 m a.s.l. on a marine terrace
at the foot of Glavnyy Val. The snow was gone by 1 July. The ground is
pebbly with a mixture of small rocks and a small amount of loam.
Individual cushions and crusts of *Stereocaulon rivulorum* predominate,
covering *c*. 45% of the surface. There are also cushions of mosses,
fragments of black film, tussocks of *Phippsia algida* and single specimens
of flowering plants.

Relevé no. 37. A site near Zaliv Dezhneva with a slight slope towards
the west at an elevation of *c*. 20 m a.s.l. The ground is a mixture of stones
measuring from 10 to 40 cm, sand and gravel. Everywhere cushions of
Stereocaulon rivulorum (cover: 40%) are scattered together with many
black crusts; some groups of *Polytrichum alpinum* stalks, individual
tussocks of *Phippsia algida* and single representatives of herbs can also be
seen. The general cover amounts to 66%.

Relevé no. 38. The site is similar to that of relevé no. 37 but wetter. The
snow reached 45 cm in thickness here but was gone by 1 July. The
substrate consists of rocks together with gravel, sand and loam. There is
much black crust and patches of white crustose lichens. Everywhere
cushions and small pieces of *Stereocaulon rivulorum* can be seen. Here and
there small groups of shoots of *Polytrichum alpinum* appear as well as a
few herbs and some *Phippsia algida*. The general cover is 50%.

Relevé no. 39. A terrace on a level, slightly rolling surface with the lower
portion of a slope on Glavnyy Val at an altitude of *c*. 15 m a.s.l. The
substrate consists of pebbles together with sand and gravel. The snow was
gone by 10 July. *Stereocaulon rivulorum* (cover: 50%) forms an almost
pure mat distributed over the pebbly ground. Small patches of black crust
are widely scattered. Mosses of low stature are hidden among stones and
pebbles and are not clearly visible in the general aspect. The same applies
to the flowering plants.

Relevé no. 40. In the same area as relevé no. 39. The snow was gone by
10 July. On the level, sandy-pebbly surface mats of *Stereocaulon rivulorum*, 50–100 cm wide and 60–200 cm long, are scattered and cover *c*.
50% of the surface. Within them are scattered black (and occasionally

grey) surface crusts, a small amount of mosses and individual specimens of flowering plants.

Relevé no. 47. A gentle slope at 20 m elevation. The snow was 50–60 cm deep but was gone by 15 July. The ground consists of small rocks with some addition of loam and many stones, measuring from 20 × 30 up to 60 × 80 cm. Basically the vegetation is represented by scattered cushions and small mats of *Stereocaulon rivulorum* and by black crusts. The presence of mosses and flowering plants is insignificant. The general cover amounts to 55%.

Relevé no. 55. A site with a level surface within a shallow basin at an elevation of 25 m a.s.l. near Zaliv Dezhneva. The ground is sandy and gravelly with a mixture of loam. The snow reached 70 cm in thickness but had disappeared by 15 July. Disjunct cushions and mats of *Stereocaulon rivulorum* (coverage 30%) and black crust predominate. Tussocks of *Phippsia algida*, individual small hummocks of *Cerastium regelii* ssp. *caespitosum* and some shoots of *Polytrichum alpinum* also occur.

Relevé no. 63. A very gentle (0.5–1°) slope down towards a rather shallow lake basin in the central part of the Tsentral'naya Susha. The snow was gone by 1 July. There are many small mats of *Stereocaulon rivulorum*, which cover 25% of the surface, and patches of black film on the small rocks mixed with gravel and loam; the plants grow both on the stone outcrops and on patches of fine soil. Cushions of mosses, associated with crustose lichens, can be seen. Tiny specimens of flowering plants are widely scattered.

Relevé no. 64. The lower portion of a south-facing slope of Glavnyy Val. The snow was gone by 1 July. The ground is sandy-pebbly. Among the pebbles, cushions of *Stereocaulon rivulorum*, patches of mosses and individual specimens of flowering plants are scattered. The general cover is 35%.

Relevé no. 65. The north-western part of the Tsentral'naya Susha; a gentle (*c.* 5°) slope at an elevation of 20 m a.s.l. The snow was gone by 15 July. The ground consists of small rocks. Patches of *Stereocaulon rivulorum* (coverage *c.* 35%) predominate. Groups of stalks of *Polytrichum fragile* can be seen as well as individual, small cushions of other mosses and separate tufts of flowering plants.

Relevé no. 69. A small site in a brook valley at an elevation of 15 m a.s.l. near Zaliv Dezhneva. The snow was 45–60 cm deep but gone by 1 July. The substrate is sandy-pebbly. There are patches of *Stereocaulon rivulorum* (coverage 20%) and black crusts. Here and there individual tufts of herbs, cushions of mosses and small tussocks of *Phippsia algida* occur.

Table 16. *Floristic composition and estimated plant cover (in %) in open aggregations belongings to type no. 7 (Comitium no. 7)[a]*

Species	61	62	66
Grasses	+	−	+
Phippsia algida	+	−	+
Herbs	+	+	+
Cerastium arcticum	+	−	−
C. regellii ssp. *caespitosum*	−	−	+
Cochlearia groenlandica	+	−	−
Draba oblongata	−	−	+
D. pauciflora	+	−	−
Papaver polare	+	−	−
Saxifraga cernua	−	+	+
S. hyperborea	−	+	+
Stellaria edwardsii	+	−	+
Mosses	3	+	5
Andreaea rupestris	−	−	+
Bryum rutilans	1	−	−
Dicranoweisia crispula	−	+	1
Mnium rugicum	−	−	+
Pohlia cruda	1	+	+
P. nutans	−	−	+
Polytrichum alpinum	1	+	2
P. fragile	−	−	1
Psilopilum cavifolium f. *nanum*	−	−	+
Timmia austriaca var. *arctica*	−	−	+
Tortula ruralis	+	−	−
Liverworts			
Cephaloziella arctica	*	*	*
Gymnomitrium concinnatum	*	−	−
Fruticose lichens	2	1	2
Alectoria ochroleuca	−	+	−
Cetraria cucullata	−	+	−
C. nivalis	−	+	+
Stereocaulon rivulorum	2	1	2
Usnea sulphurea	−	+	−
Tubiform lichens	−	+	+
Cladonia pyxidata (squamous primary thallus)	−	−	+
Thamnolia vermicularis	−	+	−
Crustose lichens			
Candellariella cf. *vitellina*	−	−	*
Collema ceraniscum	*	(*)	(*)
Lecanora campestris	*	*	−
Lecidea sp.	−	*	*
L. dicksonii	−	*	*
Ochrolechia sp.	(*)	*	*
Pertusaria sp.	*	(*)	*
Cover of light-coloured crustose lichens	1	2	2

Table 16 (*cont.*)

Species	61	62	66
(on remnants of plants)			
Cover of a black crust of crustose lichens, blue-green phycobionts and liverworts	3	2	3
Cover of epilithic lichens	+	1	2
Total plant cover	10	5	12

Note: [a] For an explanation of the figures and symbols see Table 7.

Open aggregations of type no. 7 (comitium no. 7). This consists of semi-aggregations with very few species on stony ground with participation of epilithic lichen species, *Stereocaulon rivulorum* and small-size mosses (Table 16).

The general cover of these semi-aggregations amounts to 5–12%. The snow was gone by 11–15 July. A small number of epilithic lichen species are found on the stones: crustose lichens (*Lecidea dicksonii*, etc.), individual specimens of *Usnea sulphurea* and the moss *Andreaea rupestris* also occur. The mosses are few in number; characteristically there are small forms of *Pohlia* and *Bryum* spp. and individual shoots of *Polytrichum alpinum* and *P. fragile* growing on the fine soil here and there. The herb cover is <1% with fewer species than in *comitium* no. 6, although still some individual specimens of poppies, *Cerastium arcticum* and a few other species are met with. These are not present where the snow melts at a later date. Such semi-aggregations are frequently met with and are distributed over roughly 15% of the Tsentral'naya Susha area.

Relevé no. 61. A level site at an elevation of 20 m a.s.l. with a slight slope towards the east near Zalive Dezhneva. The snow measured 65–75 cm deep and was gone by 10 July. The ground consists of a mixture of gravel, sand and rocks on which cushions of *Stereocaulon rivulorum*, individual specimens of *Phippsia algida* and some species of herbs, fragments of black film as well as some small patches, composed of crustose lichens and small-size mosses together with individual specimens of flowering plants, are unevenly scattered about. The general cover amounts to 10%.

Relevé no. 62. A stone net on a very gentle slope in the central part of the Tsentral'naya Susha. The snow was gone by 1 July. The stones are almost bare; between them on patches of fine soil and on dead cushions of mosses there are white crustose lichens and black crusts. *Stereocaulon rivulorum* is met with in some quantity; other crustose lichens are very rare. There were only two species of flowering plants. About 5% of the surface is covered by vegetation.

Relevé no. 66. A stone net on a *c.* 5° steep slope. The snow had disappeared by 15 July. The stones measure mostly 20 × 30, 35 × 50 cm, but can be as much as 120 cm across, and are almost bare; small fragments of vegetation (cover: 12%) are crowded into the intervals between them.

Open aggregations of type no. 8 (comitium no. 8). This consists of semi-aggregations with poorly developed mosses and a relatively large amount of *Phippsia algida* (Table 17).

Semi-aggregations of this type are met with on loams broken into polygons and mixed with gravel, small rocks or pebbles and also on a mixture of large stones and fine soil. They occur in habitats where the snow disappears between 1 and 15 July and are met with on less than 1% of the surface of the Tsentral'naya Susha. Essentially they can be distinguished by the poor development of the mosses and the predominance of crustose lichens and by the relative abundance of *Phippsia algida*, the cover of which in some cases reaches up to 10%. In wet habitats this species does not form the usual kind of tussocks but rather it forms miniature hummocks (*subcomitium 8″*). The small number of mosses and the fact that there is no *Ditrichum flexicaule* distinguish the semi-aggregations of the type described from the groups in comitium no. 4, where there is also much *Phippsia algida* (see Table, 13, relevé 8).

Subtype (subcomitium) 8′. This consists of semi-aggregations with tussocks of *Phippsia algida*.

Relevé no. 17. A level site at an altitude of 25 m in the north-western portion of the Tsentral'naya Susha. The snow was gone by 1 July. Loam with a slight addition of small stones is broken by fissures into polygons (Figs. 34, 35). Along the edges where the fissures intersect there is, locally, a small quantity of mosses. On the polygons there are patches, 5–20 cm in diameter, formed mainly of crustose lichens; here and there shoots of *Polytrichum fragile*, small patches of *Bryum teres*, cushions of *Stereocaulon rivulorum* and some species of flowering plants can be found. There are many tiny, mostly sterile tussocks of *Phippsia algida*. The general cover is 25%.

Relevé no. 50. Very gently sloping site, facing south, in the central part of the Tsentral'naya Susha at an elevation of *c.* 20 m a.s.l. The depth of the snow amounted to 60 cm but it was gone by 15 July. The substrate consists of fine soil (small rocks with an admixture of loam), mixed with a small quantity of blocks measuring 25 × 30 cm but not forming a well-defined stone net. Black crusts and cushions of *Stereocaulon rivulorum* predomi-

Table 17. *Floristic composition and estimated plant cover (in %) in open aggregations belonging to type no. 8 (Comitium no. 8, subcomitia nos. 8′ and 8′′)*[a]

Species	8′		8′′	
	17	50	27	52
Grasses	5	5	10	5
Phippsia algida (turf)	5	5	−	−
Phippsia algida (tussocks)	−	−	10	5
Herbs	4	+	1	+
Cerastium arcticum	+	−	−	−
C. regelii ssp. caespitosum	3	−	−	−
Draba oblongata	+	−	+	−
Saxifraga cernua	+	+	−	+
S. hyperborea	−	+	+	−
Stellaria edwardsii	−	−	+	−
Mosses	2	6	2	7
Bryum rutilans	−	−	1	3
B. teres	1	−	−	−
Dicranoweisia crispula	−	−	−	+
Hygrohypnum polare	−	5	−	3
Pohlia cruda	−	−	1	1
Polytrichum alpestre	−	−	−	+
P. alpinum	−	1	+	−
P. fragile	1	−	−	−
Rhacomitrium canescens	−	−	−	+
Schistidium gracile	+	−	−	−
Liverworts				
Cephaloziella arctica	*	*	*	*
Lophozia alpestris	−	*	−	*
L. excisa	−	*	−	−
Scapania calcicola	−	−	−	*
Tritomaria scitula	−	−	−	*
Fruticose lichens	1	15	7	2
Cetraria delisei	+	−	−	−
Stereocaulon rivulorum	1	15	7	2
Tubiform lichens	+	−	−	−
Cladonia pocillum (horizontal thallus)	+	−	−	−
C. pyxidata	+	−	−	−
Crustose lichens				
Buellia sp.	*	−	−	−
Caloplaca subolivacea	*	−	−	−
Collema ceraniscum	*	(*)	(*)	(*)
Ochrolechia frigida	*	(*)	*	*
O. tartarea	*	−	(*)	(*)
Pertusaria sp.	*	(*)	*	(*)
Cover of light-coloured crustose lichens	10	3	1	2
Cover of a black crust comprising crustose lichens, blue-green phycobionts and participation of liverworts and 'black film' consisting mainly of liverworts	10	20	5	15
Total plant cover	25	50	25	30

Note: [a] For an explanation of the figures and symbols see Table 7.

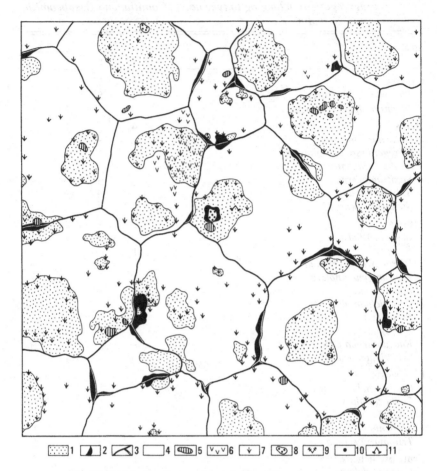

Fig. 34. Semi-aggregations on loam, broken into polygons, on Zemlya Aleksandra. The snow was gone by the beginning of July (Relevé no. 17, Table 17). Area of plot: 1 m². *1*, crust, formed by crustose lichens together with algae and, locally, tiny mosses (*Pohlia cruda*, individual shoots of *Polytrichum fragile*, etc.); *2*, a thin moss sward with predominantly *Bryum teres* in the fissures; *3*, fissures; *4*, bare, loamy soil with a small mixture of stones; *5*, small cushions of *Stereocaulon rivulorum*. *6*, *Polytrichum fragile* (individual shoots); *7*, fragments of mostly sterile specimens of *Phippsia algida*; *8*, sterile and partly dead cushions of *Cerastium regelii* ssp. *caespitosum*; *9*, *Cerastium arcticum*; *10*, *Draba oblongata*; *11*, *Cladonia pyxidata*.

nate in the vegetation, occupying 50% of the plot. Individual tussocks of *Phippsia algida* (cover: 5%) are widely scattered here and there.

Subtype (subcomitium) no. 8″. This consists of semi-aggregations with small hummocks of *Phippsia algida*.

Relevé no. 27. A site at an elevation of 10 m a.s.l. on a marine terrace at the foot of the Glavnyy Val. The snow had disappeared by the beginning of July. The ground is wet and consists of small pebbles mixed with sand and clayey loam. Scattered over this ground are disorderly round and humped hummocks of *Phippsia algida*, 5–15 cm in diameter (predominantly 8–10 cm) and 3–5 cm tall (single individuals reach up to 7 cm tall). These hummocks occupy *c*. 10% of the surface. Between them there is bare ground on which cushions of *Stereocaulon rivulorum* and clumps of mosses are met with. Old hummocks, 7–10 years old, which are dead at the centre or at one of the edges, are overgrown by mosses or covered by crustose lichens or black film. The general plant cover is 25%.

Relevé no. 52. A site in the central portion of Tsentral'naya Susha at an altitude of 15 m a.s.l. The snow was gone by 15 July. The level surface

Fig. 35. Semi-aggregation on loam broken into polygons (relevé no. 17, Table 17, Fig. 34) on Zemlya Aleksandra (photo.: the author).

consists of wet, clayey loam (the alluvium of an ancient lagoon) and is disrupted into polygons 15 × 24 to 30 × 40 cm in size. Along the fissures small mosses are met with. On the polygons, mostly near the intersections between the fissures, there are fragments of slightly wrinkled black film, sometimes pierced by small mosses, at other times not. Some cushions of *Stereocaulon rivulorum* were observed. Everywhere miniature hummocks (2–10 cm in diameter) of *Phippsia algida* are scattered. These are barren; a portion of them show various stages of dying off. The cover of *Phippsia algida* is 5%, that of the vegetation in general is 30%.

Open aggregations of type no. 9 (comitium no. 9). This consists of very impoverished fragments of black film, small mosses and *Stereocaulon rivulorum* on sandy-gravelly and fine mineral substrates (Table 18).

To this type belong semi-aggregations developed on substrates similar to those of *comitium* no. 6, but in this case the snow disappeared at an even later date, i.e., between 15 and 20 July. These semi-aggregations are met with over *c.* 5% of the area covered by the Tsentral'naya Susha, especially on low-lying surfaces of terraces and the bottoms of ancient lagoons, at the foot of deluvial shelves and on slopes with a thick snow cover. The cover of plants amounts to 3–17%, the majority of which is in the form of fragments of black film. There are very few mosses, and these are mostly species included in the *Stereocaulon rivulorum* eco-group (see Table 5). There are only four species of flowering plants, and these do not include poppies. The cover is of the same order as that of *comitium* no. 7 and differs only in species composition: in *comitium* no. 7 rupestral elements occur. As a consequence of the earlier disappearance of the snow cover, the species composition is more diversified within *comitium* no. 7.

Relevé no. 41. A marine, alluvial terrace at an elevation of 10 m a.s.l. where the snow had disappeared by 15 July. Among the pebbles mixed with sand and gravel are scattered hummocks of *Phippsia algida*, 3–12 cm in diameter; their cover being <1%. The hummocks are sterile, in part dead at the centre or on one side. Individual cushions of *Stereocaulon rivulorum* occur. Here and there small mosses or very small specimens of herbs are also met with. The general cover is 3%.

Relevé no. 48. The central portion of the Tsentral'naya Susha; a slope (*c.* 5–7°) exposed towards the south-west. The snow depth was 70 cm; the snow had disappeared by 15 July. The ground consists of fine skeletal soil; larger rocks stretch across the slope in not very distinct rows. Individual cushions of *Stereocaulon rivulorum*, individual tussocks of *Phippsia algida* and black crust are met with; here and there shoots of *Pogonatum urnigerum* occur. Plants cover 5% of the surface of this stand.

Relevé no. 51. The same area as relevé no. 48. A site at a lower level. The snow cover was 75 cm deep; it had disappeared by 20 July. On small, wet

Table 18. *Floristic composition and estimated plant cover (in %) in open aggregations belonging to type no. 9 (Comitium no. 9)*[a]

Species	41	48	51	56	60
Grasses	1	+	+	+	1
Phippsia algida	1	+	+	+	1
Herbs	+	+	+	+	+
Draba oblongata	+	–	–	–	–
Saxifraga cernua	+	+	+	+	+
S. hyperborea	+	+	–	+	+
Mosses	+	1	3	2	2
Bryum pallescens	+	–	–	–	–
B. tortifolium	–	–	+	–	–
Dicranoweisia crispula	–	+	–	–	+
Drepanocladus uncinatus	–	–	–	+	+
Hygrohypnum polare	+	–	2	–	+
Pogonatum urnigerum var. *subintegrifolium*	–	+	–	–	–
Pohlia cruda	–	+	+	–	–
P. drummondii	+	–	–	–	–
P. obtusifolia	–	–	–	1	–
Polytrichum alpestre	–	–	–	1	–
P. alpinum	+	–	–	–	–
P. fragile	–	1	+	–	1
Rhacomitrium canescens	–	+	–	–	+
Seligera polaris	–	–	–	+	–
Liverworts					
Blepharostoma trichophyllum	*	–	–	–	–
Cephaloziella arctica	*	*	*	*	*
Lophozia excisa	–	*	–	–	–
Scapania calcicola	–	*	–	–	–
Tritomaria scitula	*	*	–	–	–
Fruticose lichens	1	1	+	3	1
Stereocaulon rivulorum	1	1	+	3	1
Tubiform lichens	+	–	–	–	–
Cladonia pyxidata (squamous primary thallus)	+	–	–	–	–
Crustose lichens					
Ochrolechia frigida	–	*	–	–	–
Cover of light-coloured crustose lichens	+	1	–	2	1
Cover of fragments of 'black film'	1	2	3	10	7
Total plant cover	3	5	6	17	12

Note: [a] For an explanation of the figures and symbols see Table 7.

rocks, mixed with sand, rows of mosses are unevenly scattered together with cushions of *Stereocaulon rivulorum* and black crust. On this site there are just a few individual specimens of flowering plants, i.e., *Phippsia algida* and *Saxifraga cernua*. The general cover is 6%.

Relevé no. 56. A habitat near Zaliv Dezhneva, on a low-lying part of a gentle slope. The snow cover reached about 1 m in depth and had disappeared by 20 July. The ground consists of a sandy loamy and a fine mineral deluvium. Patches of black film (cover: 10%), some mosses and individuals of three species of flowering plants (including *Phippsia algida*) occur here. The general cover is 17%.

Relevé no. 60. A site on the central portion of the Tsentral'naya Susha with a wet, low-lying habitat. The depth of the snow cover was *c*. 1 m but the snow was gone by 20 July. The substrate consists of small stone blocks, gravel and sand mixed with loam. The vegetation consists of small, unevenly scattered mosses associated with crustose lichens. Here and there very small tussocks of *Phippsia algida* can be seen; among the herbs only individuals of *Saxifraga cernua* are met with. The general cover amounts to 12%.

Open aggregations of type no. 10 (comitium no. 10). This consists of very impoverished semi-aggregations of small patches of mosses with fragments of black film and no flowering plants (Table 19).

Semi-aggregations of this type are met with on wet clayey loam broken into polygons (on sites of ancient lagoons on low marine terraces) and occupy less than 1% of the Tsentral'naya Susha area.

Table 19. *Floristic composition and estimated plant cover (in %) in open aggregations belonging to type no. 10 (Comitium no. 10)*[a]

Species	53
Mosses	9
Hygrohypnum polare	9
Oncophorus virens	+
Liverworts	
Cephaloziella arctica	*
Lophozia alpestris	*
Fruticose lichens	+
Stereocaulon rivulorum	+
Cover of fragments of 'black film'	1
Total plant cover	10

Note: [a] For an explanation of the figures and symbols see Table 7.

Relevé no. 53. A level site with wet, sticky, clayey loam, shaped into small polygons bordered by fissures. The snow cover was more than 1 m deep but had disappeared by 20 July. Small amounts of mosses – among them *Hygrohypnum polare* predominates – form groups of tiny patches mainly at the intersections of the fissures. Here and there fragments of black film and individual, very small cushions of *Stereocaulon rivulorum* can be seen. There are no flowering plants. The general cover amounts to 10%.

Open aggregations of type no. 11 (comitium no. 11). This consists of aggregations of individual specimens belonging to the *Stereocaulon rivu-*

Table 20. *Floristic composition and estimated plant cover (in %) in open aggregations belonging to type no. 11 (Comitium no. 11, subcomitia nos. 11′ and 11″)*[a]

Species	11′			11″
	49	59	67	57
Grasses	+	+	+	−
Phippsia algida	+	+	+	−
Herbs	+	−	+	−
Saxifraga cernua	+	−	+	−
S. hyperborea	+	−	+	−
Mosses	+	+	+	+
Dicranoweisia crispula	−	+	+	−
Hygrohypnum polare	+	+	+	−
Myurella julacea	−	−	−	+
Pogonatum urnigerum var. *subintegrifolium*	−	+	−	−
Pohlia nutans	+	−	−	−
P. obtusifolia	−	−	+	+
Polytrichum alpestre	−	−	+	−
Rhacomitrium canescens	−	+	−	−
Seligera polaris	−	−	+	+
Liverworts				
Cephaloziella arctica	−	*	−	−
Fruticose lichen	+	+	+	+
Cetraria delisei	−	−	−	+
Stereocaulon rivulorum	+	+	+	+
Crustose lichens				
Ochrolechia frigida	−	*	−	−
Total plant cover	+	+	+	+

Note: [a]For an explanation of the figures and symbols see Table 7.

lorum eco-group; flowering plants occur as individual specimens only or are entirely absent (Table 20).

The type in question comprises aggregations of extreme poverty belonging to the category of open aggregates. No fragments of phytocoenoses are met with, not even in the form of patches of black film; only specimens growing individually are encountered. They cover a fraction of 1% of the surface. There are frequently no flowering plants at all. Such aggregates can be seen on habitats where the snow persists until 20 or 25 July and they occupy 35% of the surface of the Tsentral'naya Susha.

Subtype (subcomitium) no. 11'. This includes individual specimens of flowering plants.

Relevé no. 49. A habitat on the central part of the Tsentral'naya Susha; an almost horizontal surface below a gentle slope. The thickness of the snow cover was 95 cm but the snow was gone by 20 July. Among the stones small patches of mosses can be seen here and there; extremely rarely cushions of *Stereocaulon rivulorum* (0.5–1.5 cm in diameter) and tiny specimens of *Phippsia algida, Saxifraga cernua* and *S. hyperborea* are met with.

Relevé no. 59. A conglomeration of stones on a wet, low-lying habitat. The snow depth amounted to 1 m and persisted until 20 July. The stones are bare; here and there individual patches of mosses and single individuals of *Stereocaulon rivulorum* are widely scattered; only a single individual of *Phippsia algida* was observed.

Relevé no. 67. A stony substrate on the lower portion of a slope; the snow was gone by 20 July. At first glance no plants can be located (Fig. 36); only on closer scrutiny can individual cryptogams belonging to the *Stereocaulon rivulorum* eco-group and very tiny specimens of *Phippsia algida, Saxifraga cernua* and *S. hyperborea* be found.

Subtype (subcomitium) no. 11". This has no flowering plants at all.

Relevé no. 57. A gently sloping site near Zaliv Dezhneva. The snow cover was more than 1 m deep but had disappeared by 25 July. The substrate is sandy-loamy-gravel. Here and there cushions of *Stereocaulon rivulorum* and tiny groups of moss-shoots are encountered. There are no flowering plants.

Open aggregations of type no. 12 (comitium no. 12). This consists of vegetation favouring mossy bog-like habitats (Table 21).

These semi-aggregations have developed on very wet habitats, constantly moist. Mosses saturated with water cover 30–65% of the surface in the form of rows and strips. Flowering plants are met with only as single individuals.

Relevé no. 43. A site in a stream bed in wet mineral ground. The snow disappeared between 10 and 15 July. Mosses saturated with water and covering 40% of the surface form rows and strips 15–150 cm wide among the stones. Among the flowering plants only tiny, single specimens of *Phippsia algida* occur.

Relevé no. 44. Aggregations on the sandy-pebbly alluvium of a stream bed appear in the form of delicately rose-coloured mosses (species of *Orthothecium* and *Bryum*) saturated with water, which grow among the stones and cover *c.* 30% of the surface. Here and there tiny, single specimens of *Phippsia algida* occur. The snow disappeared between 10 and 15 July.

Fig. 36. Stony ground with individual small specimens of badly stunted plants among the stones on a habitat with late-melting snow, on Zemlya Aleksandra (photo.: the author).

Relevé no. 58. A stony-clayey stream bed on which brightly coloured green, golden and rusty cushions of mosses grow interspersed with single tussocks of small-size *Phippsia algida*, very small rosettes of *Cochlearia groenlandica* and very tiny specimens of *Saxifraga cernua*. In small pools, dark and curly convoluted sheets of *Nostoc commune* can be seen. The snow had disappeared by 1 July. The plant cover amounted to 45%.

In Table 22 a summary is presented of the general cover of the vegetation, the participation of the flowering plants in the types of

Table 21. *Floristic composition and estimated plant cover (in %) in open aggregations belonging to type no. 12 (Comitium no. 12)*[a]

Species	43	44	58
Grasses	+	+	+
Phippsia algida	+	+	+
Herbs	−	−	+
Cochlearia groenlandica	−	−	+
Saxifraga cernua	−	−	+
Mosses	40	30	45
Bryum rutilans	−	−	5
B. teres	−	+	−
B. tortifolium	−	5	−
Calliergon sarmentosum	−	−	+
Campylium stellatum	25	−	7
Drepanocladus uncinatus	10	−	−
Hygrohypnum polare	4	−	2
Orthothecium chryseum	1	25	25
O. strictum	−	+	−
Philonotis tomentella	−	−	+
Liverworts			
Blepharostoma trichophyllum	*	*	*
Cephaloziella arctica	−	−	*
Fruticose lichens	−	+	+
Stereocaulon rivulorum	−	+	+
Crustose lichens			
Ochrolechia cf. *tartarea*	−	−	*
Cover of light-coloured crustose lichens	−	−	+
Cover of fragments of 'black film', consisting mainly of liverworts	−	−	1
Total plant cover	40	30	46

Note: [a] For an explanation of the figures and symbols see Table 7.

aggregations described and the extent of their occurrence on Zemlya Aleksandra. The ecological ordination of the typological units is illustrated in Fig. 19.

The vegetation on Zemlya Aleksandra is characteristic of the northern belt of the arctic polar deserts. Safronova (1983), visiting the southern islands of the archipelago, i.e., Ostrova Meybel and Gukera (see Fig. 23), in 1979, described the vegetation there as belonging to the southern belt of the arctic polar deserts: due to the presence of locally favourable conditions the vegetation displays extrazonal associations similar to those of the arctic tundra (Aleksandra, 1977a). On Ostrov Gukera (and also on Ostrov Kheysa) *Salix polaris* occasionally occurs in some of these associations.

The vegetation is more satisfactorily developed on the southern islands of the archipelago of Zemlya Frantsa-Iosifa, apparently due to the fact that the western branch of the relatively warm western Novaya Zemlya current flows a short distance from them, bringing warm water with it from the Gulf stream while the northern, western and eastern islands are surrounded by the cold waters of the transatlantic current, which carries with it ice from the arctic basin.

Ostrov Victoriya

This small island (surface area 5.3 km^2 only), which is situated within the northern belt of the arctic polar deserts, is the most westerly of the Soviet arctic islands. It is almost completely covered by a dome of ice. On the small piece of land which is free of ice, hardly any vegetation has developed. However, some information has been published by Govorukha (1970b): 'Due to the extreme "youth" of the land, free of ice, the vegetation on Ostrov Victoriya is represented almost exclusively by a few species of mosses, lichens and ground-covering algae.' There was a species that was new to science in the collection gathered by Govorukha in 1961, i.e., *Gasparrinia arctica* Golubk. et Savicz.

Like Zemlya Frantsa-Iosifa, Ostrov Victoriya has a maritime, cryohumid climate. In June the mean temperature is −1.2°C, in July it is 0.7°C, in August it is 0.4 °C and in September it is −1.5°C. The yearly amount of precipitation amounts to c. 250 mm. Frequent fogs, a high relative humidity, much cloudiness and strong and frequent winds are characteristic.

The northern part of Novaya Zemlya

The northernmost portion of Novaya Zemlya falls within the polar deserts (Fig. 1). The major part of this territory is occupied by an ice

Table 22. Types of nanocomplexes and open aggregations on the island of Zemlya Aleksandra

Type or no.	Occurrence (in %) of the area of Tsentral naya Susha	Dates by which the snow had disappeared completely	Aggregations	General plant cover (in %) (mean of the relevés)	Participation of flowering plants — Mean of each relevé		Flowering plants recorded from all the relevés[a]	No. of relevés
					Coverage (%)	No. of species		
Nanocomplexes								
A	<1	15–22 June	Nanocomplexes with a polygonal-netlike distribution in a sward consisting of mosses and fruticose lichens with a significant participation of species belonging to the *Cetraria cucullata* eco-group on a mixture of stones and fine soil	85	3	10	Ca, Cb, Cg, Cr, Dm, Do, Da, Lc, Pab, Pal, Ph, Sc, Sca, Se, Sh, So	10
B	1	15–22 June	Lichen-dominated nanocomplexes with a significant participation of species belonging to the *Cetraria cucullata* eco-group on sandy-pebbly and gravelly substrates with small stones	85	5	7	Ca, Cb, Cg, Do, Dp, Ds, Lc, Pab, Ph, Pp, Sc, Se, Sh	3
C	<1	15 June	Moss–lichen-dominated nanocomplexes on loamy plakor (zonal aggregations)	85	3	6	Ca, Cr, Do, Ph, Pp, Sh	1
D	<1	1 July	Hummocks of *Phippsia algida* in a black film on moist sandy-pebbly substrates	95	9	5	Dm, Dp, Ph, Pp, Sc, Sh	1
E	<1	1 July	A black film covering wet clayey loam	90	1	3	Ph, Sc, Sh	1
Open aggregations (*comitia*)								
1	2	10–22 June	Lichen (*Umbilicaria, Usnea,* etc.)-dominated semi-aggregations on stone fields	70	<1	2	Ca, Cg, Cr, Dm, Do, Ds, Ph, Pp, Sc, Sh	7
2	<1	22 June	Lichen-moss (*Rhacomitrium lanuginosum*)-dominated semi-aggregations on stone fields	55	1	3	Cr, Dp, Ph	1
3	1	End of May to 10 June	Semi-aggregations consisting of a lichen crust in sandy-pebbly and sandy-gravelly habitats with small blocks and a scanty snow cover	60	1	8	Ca, Cg, Do, Ds, Mr, Pab, Ph, Pp, Sc, Scs, Se, Sh, Sn	5
4	1	22 June	Semi-aggregations consisting of moss (*Ditrichum flexicaule,* etc.) – lichen strips and swards on loam broken into polygons	40	6	8	Ca, Cg, Cr, Do, Dp, Ds, Ph, Sc, Se, Sh	2
5	<1	22 June	Semi-aggregations consisting of fragments of black film and moss–lichen sward with *Cetraria delisei* on moist stony substrates	50	2	6	Cg, Cr, Do, Dp, Ds, Ph, Pp, Sc, Se, Sh, So	4
6	3	1–15 July	Semi-aggregations consisting of cushions and swards of *Stereocaulon rivulorum* on sandy-pebbly and gravelly blocky substrates	60	2	8	Ca, Cb, Cg, Dm, Do, Dp, Ds, Ph, Pp, Sc, Se, Sh	11

No.		Date	Description				Species[a]	
7	15	1–15 July	Very impoverished semi-aggregations on stony substrates with participation of epilithic species such as *Stereocaulon rivulorum* and small-size mosses (*Pohlia* spp., etc.)	10	<1	5	Ca, Cg, Cr, Do, Dp, Ph, Pp, Sc, Se, Sh	3
8	1	1–15 July	Semi-aggregations with a comparatively large amount of turf and hummocks of *Phippsia algida* and poorly developed mosses	30	7	3	Ca, Cr, Do, Ph, Sc, Se, Sh	4
9	5	15–20 July	Very impoverished semi-aggregations consisting of fragments of black film, small mosses and *Stereocaulon rivulorum* on substrates that are sandy-pebbly with small stone blocks	5	<1	3	Do, Ph, Sc, Sh	5
10	1	20 July	Very impoverished semi-aggregations consisting of small patches of mosses (*Hygrohypnum polare*, etc.) and fragments of black film on moist polygonal loam, completely lacking flowering plants	10	—	—	—	1
11	35	20–25 July	Aggregations consisting of various individuals from the *Stereocaulon rivulorum* eco-group; flowering plants occur individually or not at all	<1	<1	1	Ph, Sc, Sh	7
12	<1	1–15 July	Mossy bog-like habitats	40	<1	2	Cg, Ph, Sc	3
13[b]	25	After 25 July	Localities of non-glaciated land, devoid of vegetation	—	—	—	—	—
14[b]	10		Lakes or permanent snowfields	—	—	—	—	—

Notes: [a] Ca, *Cerastium arcticum*; Cb, *Cardamine bellidifolia*; Cg, *Cochlearia groenlandica*; Cr, *Cerastium regelii* ssp. *caespitosum*; Dm, *Draba macrocarpa*; Do, *Draba oblongata*; Dp, *D. pauciflora*; Ds, *D. subcapitata*; Lc, *Luzula confusa*; Mr, *Minuartia rubella*; Pab, *Poa abbreviata*; Pal, *P. alpigena*; Ph, *Phippsia algida*; Pp, *Papaver polare*; Sc, *Saxifraga cernua*; Scs, *S. caespitosa*; Se, *Stellaria edwardsii*; Sh, *Saxifraga hyperborea*; Sn, *S. nivalis*; So, *S. oppositifolia*.
[b] Types nos. 13 and 14 are discussed in the text.

sheet the altitude of which reaches 1000 m a.s.l. The narrow belt that is free of ice is locally reduced to nothing but widens on Mys Zhelaniya, where it becomes a ridged and rolling plain with stony morainic deposits cut by rivers and streams originating from the ice shield (Zubkov, 1934; Samoylovich, 1937). In June the mean temperature is −1.2°C, in July it is 1.7°C, in August it is 2.1°C and in September it is −0.1°C. Extremely strong winds dominate; the mean wind force of Mys Zhelaniya is 8.2 m/s. Although there is less precipitation than on Zemlya Frantsa-Iosifa, i.e., a yearly total of 189 mm, the summers are wet due to frequently occurring fogs, a high relative humidity of the air and the occurrence of trace amounts of precipitation.

No lists of the flora from this area exist, but an enumeration of the vascular plants from the northern part of the northern island of Novaya Zemlya has been published by Tolmachev (1936); it concerns the area north of latitude 75° N. The southern part of that area belongs within the arctic tundra.

On the basis of this information, the vegetation of the northern part of the area in question belongs to the southern belt of the arctic polar deserts. However, the information is very scanty and is of only a general character. Thus, Tolmachev (1936, p. 167) states: '... the vegetation cover is everywhere very poor and in many places represented by plant associations almost impossible to describe, i.e., only individual plants occur over wide stretches of land with no other vegetation.' Somewhat more detailed but still scanty information is provided by Zubkov (1934): 'The stretches of land on the north-eastern coast of the North Island ... have a scanty vegetation, consisting of scattered individual specimens but also to some extent of fragments of phytocoenoses... The lack of a snow cover in some areas and the extreme accumulation of snow in other areas result in the fact that in many cases there are extremely unfavourable conditions for the development of a vegetation. On the tops of ground moraines, on slopes facing south, and everywhere where, most of the time, the snow is blown away by strong winds, it is possible to observe individual specimens among the stones, such as *Saxifraga oppositifolia*, which are able to grow while protected on the north-facing side by the micro-relief ... In habitats where snow accumulates the vegetation takes on the character of a few fragments of phytocoenoses, although the role of the flowering plants is insignificant because of the daily melting of snow. Here mainly lichens and a few flowering plants are met with, e.g., *Cerastium regelii*, *Poa alpigena* and *Oxyria digyna*. On wet slopes with various exposure and in shallow depressions stoney polygon-formations are distributed; where on the polygons single specimens of *Saxifraga oppositifolia*, *Cerastium regelii*,

Draba alpina, etc. grow, and along the edges of the polygons, on the stone rings, some *Cetraria delisei* has developed together with crustose lichens and mosses belonging to the genus *Drepanocladus*. A bog-like type of vegetation has developed in wet habitats, in rills of melt-water and on shores with puddles where some habitats with hummocks of *Deschampsia borealis* or fragments of grass–*Hypnum* phytocoenoses are met with.'

Fig. 37. Area of distribution of *Draba barbata* Pohle (according to Hultén, 1968; *Arctic flora of the USSR*, vol. 7, 1975; Safronova, 1981b).

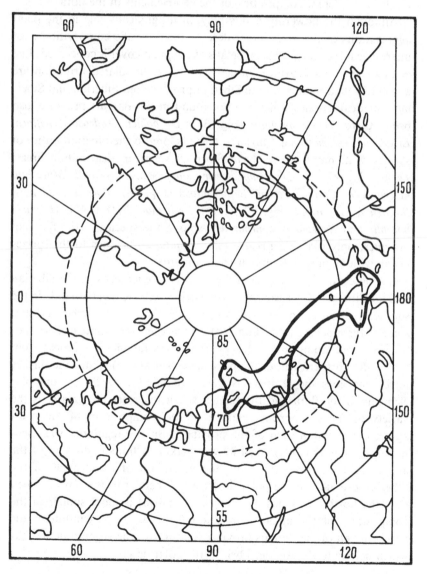

The vegetation within the Siberian province of the polar deserts

The archipelago of Severnaya Zemlya, the adjacent Mys Chelyuskin and Ostrova De Longa belong to the Siberian province of the polar deserts. The composition of the plant association differs from that of the plant associations found in the Barents province not only because of the differences in climate, which in the Siberian Province has a more continental character, but also because of different florogenic connections, which result in a different composition of the geo-elements in the flora.

Although the flowering plants play an insignificant role in the plant aggregations of the arctic polar deserts, their specific composition is of interest as far as the provincial details of the plant cover is concerned. Just as in the Barents province, species with a circumpolar distribution are absolutely predominant in the Siberian province but Siberian and Siberian-American species which are not found in the Barents province also occur there. *Draba barbata* (Fig. 37), *D. pohlei, Androsace triflora, Eritrichum villosum* and *Saussurea tilesii* belong to the Siberian group of species. Still more species have Siberian-American distribution areas: *Festuca hyperborea, Cerastium beeringianum* ssp. *bialynickii, Minuartia macrocarpa, Caltha arctica, Draba-pilosa* (Fig. 38), *Parrya nudicaulis, Saxifraga serpyllifolia* (Fig. 39), *Novosieversia glacialis* (Fig. 40), *Artemisia borealis* and *Senecio atropurpureus*. Among the species in the Siberian province, only *Cerastium regelii* ssp. *caespitosum* belongs to the species group with an amphiatlantic distribution area.

It is necessary to take note of the fact that the majority of the Siberian and the Siberian-American species occurring here are met with only in the southern belt of the polar deserts within the Siberian province, and that some of them belong to the families Primulaceae, Boraginaceae, Scrophulariaceae and Compositae; species from these families are absent from within the northern belt of the polar deserts (Table 6). Thus, the further north one goes the stronger is the predominance of the circumpolar elements in the flora and the further south one goes the greater is the variety of the components making up the composition of the plant aggregations within each of the provinces. Such a regularity was mentioned by Young (1971) when he was making a floristic division of the arctic areas. In this connection, the vegetation of those areas of the Siberian province of the polar desert which belong to the northern belt (i.e., the northern islands of Severnaya Zemlya) is more similar in the composition of the angiosperm components to the vegetation of the northern belt of the Barents province than it is to the vegetation of areas which belong to the southern belt of the polar deserts.

Since the role of flowering plants in the polar desert is always insignificant, and since cryptogams take the leading position in the composition of the plant aggregations – especially lichens (and particularly crustose ones) – the most essential characteristic appears to be the specific composition of the cryptogamic components. The composition of the ground-covering crusts of crustose lichens appears to be the most distinc-

Fig. 38. Area of distribution of *Draba pilosa* DC. (according to Hultén, 1968; *Arctic flora of the USSR*, vol. 7, 1975; Safronova, 1981b).

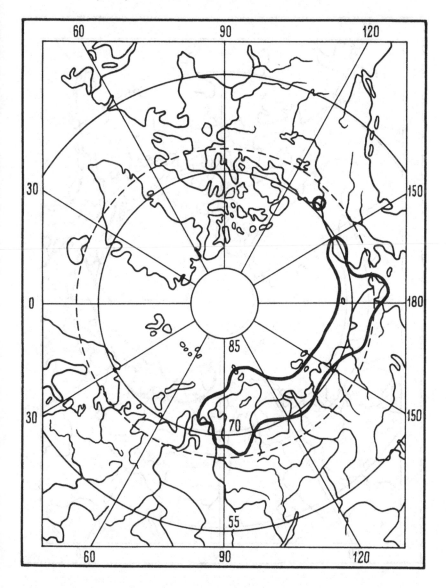

tive characteristic of the plant aggregations in the Siberian province of the polar deserts. It is practically predominant in its cover and seems to be the characteristic synusium of the type of vegetation in question.

As demonstrated by the investigation made by Matveyeva (1979) and Piyn (1979), *Toninia lobulata* is the main component of this crust in the area of Mys Chelyuskin (evidently it also occurs on Severnaya Zemlya,

Fig. 39. Area of distribution of *Saxifraga serpyllifolia* Pursh (according to Hultén, 1968).

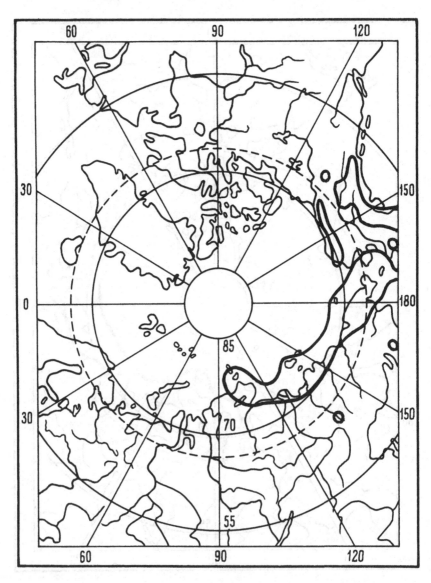

according to the observations made by Korotkevich) while in the Barents province the basic components of the surface crust under plakor conditions and conditions approaching plakor are species of *Pertusaria* and *Ochrolechia*. These observations highlight the physiognomic distinction between the plant aggregations: the grey crust of *Toninia lobulata*, with its black apothecia, is the one most widely distributed in the Mys Chelyuskin

Fig. 40. Area of distribution of *Novosieversia glacialis* (Adam) F. Bolle (according to Hultén, 1968).

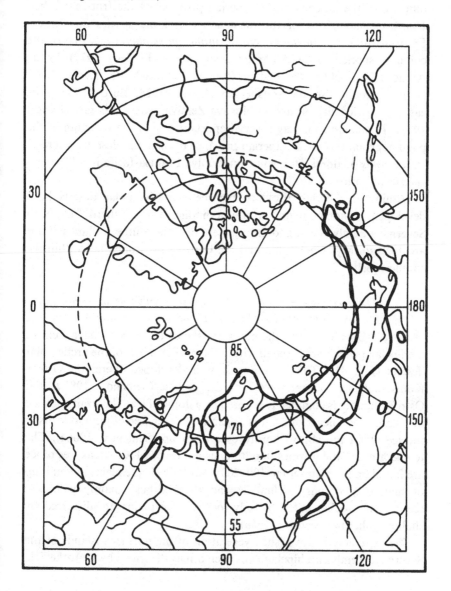

area, while on Zemlya Frantsa-Iosifa the background synusia in the plakor aggregations consist of white crustose lichens, where the pure-white patches of *Pertusaria glomerata* are associated with white *Ochrolechia frigida*, against the white background of which black spots of *Collema* and other species, forming a black crust, are scattered. Matveyeva (in Matveyeva and Chernov, 1976) based her opinion on and emphasized the importance of this essential difference between the Mys Chelyuskin and the Zemlya Frantsa-Iosifa areas. But there are also other differences: the absence in the Siberian province of the fruticose lichen *Usnea sulphurea*, one of the most active epilithic species on Zemlya Frantsa-Iosifa (*U. sulphurea*, syn. *Neuropogon sulphureus*, is one of the common species on the Antarctic continent, see Lindsay, 1977). In the Arctic it is widely distributed on Zemlya Frantsa-Iosifa (Lynge, 1931, etc.): it occurs also on Spitsbergen (Lynge, 1928; Eurola, 1971); its easternmost locality occurs on Novaya Zemlya, where it is rare (Lynge, 1928). *Thamnolia* spp. play a very important role in the structure of the plant aggregations of the Siberian province of the polar deserts, where *T. subuliformis* predominates, while on Zemlya Frantsa-Iosifa it is *T. vermicularis* and others.

Within the Siberian province two belts of the polar deserts have developed, a northern and a southern one. The northern islands of Severnaya Zemlya belong to the northern belt, the southern islands of this archipelago, Mys Chelyuskin and the Ostrova De Longa fall within the southern belt.

The northern islands of the Severnaya Zemlya archipelago

Ostrov Komsomolets, Ostrov Pioner, Ostrov Shmidta and the Ostrova Sedova comprise the northern islands of the Severnaya Zemlya archipelago which are situated within the northern belt of the arctic polar deserts (Fig. 41). The ice cover is well developed there: on Ostrov Komsomolets it occupies 69% of the surface (Semenov, 1981*a*) and Ostrov Shmidta is completely covered by a dome of ice. The strait between the islands may never open; it is apparently never free of ice and leads in the ice along the shores hardly ever open up (Semenov, 1966*a*). The temperature of the warmest month does not exceed 1°C. Areas free of ice are found on the plains up to 100–200 m altitude. These plains are built up of marine quaternary sediments and the bedrock (mainly paleozoic sedimentary rocks) outcrops only locally. The plains are dissected by shallow valleys (Semenov, 1966*b*).

The extremely impoverished vegetation of the northern islands of the Severnaya Zemlya archipelago has been briefly described by Korotkevich

(1958) and Semenov (1966*b*). On Ostrov Pioner 8 species of flowering plants were found, on Ostrov Komsomolets and Ostrova Diabazovykh 5 species were found and on Ostrova Sedova 15 species were found (see Table 6). The species belong to the families Gramineae, Caryophyllaceae, Papaveraceae, Cruciferae and Saxifragaceae. Neither willows nor any representatives of gamopetalous species or even plants of the family Rosaceae have ever been discovered on these islands.

Korotkevich (1958) has produced the most detailed account of the vegetation on Ostrov Srednyy (the middle island of Ostrova Sedova). Here

Fig. 41. Map of Severnaya Zemlya (according to Semenov, 1966*b*; Korotkevich, 1972). *1*, boundary between the northern and the southern belts of the polar deserts; *2*, glaciated areas.

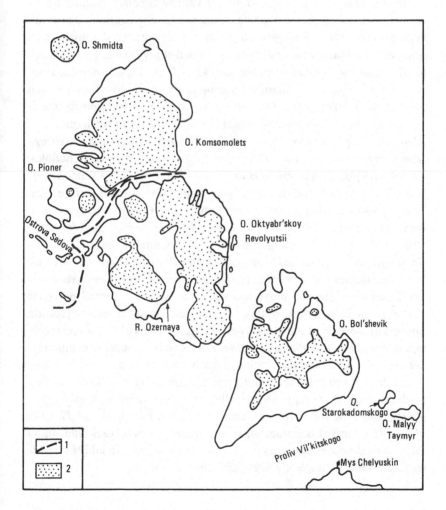

only an insignificant surface area is occupied by aggregations with a polygonal network structure (nanocomplexes) where, along the fissures separating the polygons (which are 30–50 cm in diameter), a moss–lichen sward consisting predominantly of *Cetraria delisei, C. cucullata* and crustose lichens is distributed. Among the mosses *Distichium capillaceum* and *Ditrichum flexicaule* predominate, while among the flowering plants *Saxifraga oppositifolia* dominates; the other angiosperm components, *Cerastium regelii, Draba pauciflora, Papaver polare, Phippsia algida, Saxifraga caespitosa, S. cernua* and *Stellaria edwardsii,* play an insignificant role. The general cover amounts to 50%. The centres of the polygons are either bare or covered with a surface crust of predominantly crustose lichens.

The remaining plant association on Ostrov Srednyy mentioned by Korotkevich (1958) belong to the categories of aggregations and semi-aggregations. The semi-aggregation most satisfactorily developed was observed in a shallow depression between two stony ridges. There a level, loamy surface was occupied by hummocks of *Deschampsia caespitosa* ssp. *glauca,* reaching 30 cm in diameter and up to 15 cm tall. Moss species such as *Distichum capillaceum, Orthothecium chryseum, Ditrichum flexicaule* and *Campylium zemliae* played a part in the formation of the hummocks; *Alopecurus alpinus, Cerastium regelii, Phippsia algida* and *Saxifraga cernua* were also met with. Colonies of *Nostoc* were observed in puddles. The general plant cover amounts to 70–80%. According to a statement by Korotkevich (1958), such a vegetation occupies only a small surface area. Three types of open aggregations predominate on Ostrov Srednyy: one of them, consisting mainly of lichens, especially crustose species, is found on structured ground with a large amount of stones; among the angiosperms, *Saxifraga oppositifolia, Saxifraga cernua* and *Papaver polare* are met with but other species are very rare. The second type of open aggregation has developed on a fine soil polygonal substrate, where the small polygons (20–30 cm in diameter) are outlined by fissures. There the main vegetation appears in the form of individually growing specimens of *Cerastium regelii* ssp. *caespitosum* and *Phippsia algida.* A relatively important component consists of a black surface coating. Finally, in low-lying, moist localities with a loamy substrate there are tussocks consisting of various mosses, predominantly *Bryum* spp. and, locally, some *Deschampsia caespitosa* ssp. *glauca,* and here and there very small individual hummocks of *Phippsia algida.* There are also surface algae and fragments of a black film. In all three cases the cover amounts to 5–10%, rarely to 15%. In addition, wide surfaces, almost completely bare, are found there.

The vegetation of Ostrov Golomyanogo (the most westerly island in Ostrova Sedova) is even more impoverished. Korotkevich (1958, pp. 657–8) provided the following example of it: 'Over wide areas there are only individual, very tiny plants. Among the angiosperms *Phippsia algida*, locally *Cerastium regelii* and very rarely *Stellaria edwardsii, Saxifraga oppositifolia, S. caespitosa* and *S. cernua* are met with as well as very few specimens of *Papaver polare* and *Puccinellia angustata*. Mosses (*Ditrichum flexicaule*, etc.) are rare and stunted. Fruticose lichens are practically absent, only a few patches of *Thamnolia vermicularis, Dactylina madreporensis* and *Cetraria delisei*, and locally some patches of crustose lichens can be found.' Khodacheck (1980), who visited Ostrova Sedova in 1980, also mentions their extremely poorly developed vegetation.

The southern islands of the Severnaya Zemlya archipelago

Two major islands, Ostrov Oktyabr'skoy Revolyutsii and Ostrov Bol'shevik, and the smaller islands Ostrov Malyy Taymyr and Ostrov Starokadomskogo belong to the southern islands of the Severnaya Zemlya archipelago. These islands are considerably less glaciated than the northern islands: on Ostrov Oktyabr'skoy Revolyutsii the ice cover occupies 58% of the territory, on Ostrov Bol'shevik 31%; Ostrova Starokadomskogo and Malyy Taymyr are free of glaciation (Semenov, 1981*b*). The mean July temperature on Malyy Taymyr reaches 1.3°C. In coastal areas of the larger islands it ranges between 1.3 and 1.6°C (Semenov, 1966*a*) with the exception of anomalous temperatures that arise from the effect of foehn winds (see below, p. 138); higher temperatures (caused by the same effect) also occur locally in the valleys of the central portion of Ostrov Oktyabr'skoy Revolyutsii, resulting in the development there of a richer, extrazonal type of vegetation of which more is mentioned below.

Concerning the vegetation cover on the southern islands of the Severnaya Zemlya archipelago, we can only refer to the data available from Ostrov Oktyabr'skoy Revolyutsii (Korotkevich, 1958, Safranova, 1976, 1981*b*; Khodacheck, 1980). The other islands (Bol'shevik, Malyy Taymyr and Starokadomskogo) have not yet been subjected to the geobotanical investigations.

Zonal vegetation

Vegetation approaching a zonal type occurs on the wide coastal plain on the south-western portion of Ostrov Oktyabr'skoy Revolyutsii; and has been described by Korotkevich (1958). There, on a level surface, at an altitude of *c.* 25 m a.s.l., a substrate consisting of loam with a large amount of calcareous rocks forming stone nets which delimit polygons of loam with addition of rocks, is found. The general plant cover is about 20%. The plants are distributed in the form of individual patches or strips and do not form a closed sward. Lichens play a major role, especially the crustose ones (they constitute about one half of the entire cover) and *Cetraria delisei.* Locally, there is a large amount of *Saxifraga oppositifolia,* which forms large patches. (When studying the works of Korotkevich, it is obvious that he concluded that *S. oppositifolia* should, according to Gorodkov, be considered as a dwarf shrub although this species is not a dwarf shrub but a herbaceous plant (e.g. Warming, 1912; Polozova, 1978). It is necessary to use also for the examination tables of geobotanical stand surveys in which the estimated cover of various groups of species are treated (see Korotkevich, 1958, pp. 648–9).) Other herbaceous plants grow scattered and cover only about 2% of the surface although a variety of species are represented, e.g., *Cerastium beeringianum* ssp. *bialynickii, Deschampsia caespitosa* ssp. *glauca, Minuartia rubella, Papaver polare, Phippsia algida, Saxifraga caespitosa, S. cernua* and *S. nivalis.* There are few mosses (*Distichium capillaceum, Hypnum bambergeri,* etc.), with a cover of *c.* 2%.

Such plant communities, according to Korotkevich (1958), are distributed over the southern coastal area of the island. Sites which are even more impoverished and almost entirely free of any plant cover are also met with. At the foot of hills, in habitats moistened by water from melting snow, there are often bog-like sites containing scattered clumps of mosses (*Orthothecium chryseum,* etc.) and *Deschampsia caespitosa* ssp. *glauca* (in the original paper by Korotkevich (1958) it is called *D. borealis*); locally, *Alopecurus alpinus* and *Phippsia algida* are found. Other flowering plants are rare and grow in the form of individual specimens only.

Extrazonal vegetation
Vegetation which on Ostrov Oktyabr'skoy Revolyutsii is represented by associations of a more southerly type occurs primarily in the form of different phytocoenoses. In addition, these occupy relatively large tracts of territory.

Phytocoenoses of a tundra type that have been encountered in the polar desert landscape have been described by Korotkevich (1958, p. 651) from areas of the south-eastern coastal plain of Ostrov Oktyabr'skoy Revolyutsii with well-heated habitats in the neighbourhood of the Ozernaya river (Fig. 41). Although the general cover of plants amounts to only 30% and of this 10% is accounted for by crustose lichens (mosses cover only 3%), flowering plants are well represented, covering 17%. *Saxifraga oppositi-*

folia predominates but there is also much *Salix polaris;* in addition, although in small amounts, *Cerastium beeringianum* ssp. *bialynickii, C. regelii, Draba macrocarpa, D. subcapitata, Eritrichum villosum, Minuartia rubella, Papaver polare, Saxifraga caespitosa, S. cernua, S. nivalis, Stellaria edwardsii* and *Salix reptans* are also present; in general these species are very rare on Severnaya Zemlya.

Relatively large areas with an extrazonal type of vegetation are found on the peninsulas of Parizhskaya Kommuna and Zhiloy and also in the river valleys of the central parts of the island. A portion of such a habitat, in part described by Korotkevich from a river valley in the interior of the island, can be characterized as representative of tundra (analogous to the northern type of the arctic tundra). The term 'polar semi-desert' could be applied to another part with characteristics intermediate between those of tundra and polar desert. This term, although in a different sense, has been used by some non-Soviet authors (e.g. Bliss, 1975, 1977; Bell and Bliss, 1978; Addison and Bliss, 1980).

The concept of the term 'polar semi-desert' can, perhaps, also be applied to the descriptions made by Safronova of the vegetation from Ostrov Meybel (Zemlya Frantsa-Iosifa). It could be used for plant associations where a significant proportion consists of angiosperms but where the component that is basic to the Eurasiatic arctic tundra, i.e., *Salix polaris*, is absent. The same applies also to associations on Ostrov Oktyabr'skoy Revolyutsii, described by Korotkevich from the Parizhskaya Kommuna peninsula and by Safronova (1976) and Khodacheck (1980) from the Zhiloy peninsula, in cases where *Salix polaris* is not found.

The presence or absence of the arctic–alpine dwarf shrub *Salix polaris* in the composition of the plant associations constitutes an important diagnostic characteristic for the determination of whether we are dealing with a tundra or a polar desert type of vegetation, since arctic–alpine dwarf shrubs appear as characteristic synusia of the arctic subtype of tundra-type vegetation (Aleksandrova, 1971). They can occur in associations, referred by myself to types of polar desert vegetation, but are met with (within the limits of the southern belt of the polar deserts) only in extrazonal habitats, i.e., on south-exposed slopes, in more satisfactorily heated habitats, etc.

According to the description published by Korotkevich (1958), the vegetation of the Parizhskaya Kommuna peninsula can be considered to come in the category of 'arctic semi-deserts' since there, on plakor habitats among the angiosperms that are satisfactorily developed, *Salix polaris* appears in the composition of plant associations only on the south-facing slopes that are best heated. As that author states, a vegetation of a polygonal network type (polygons 30–60 cm in diameter) with a general cover amounting to *c*. 65% has developed on a narrow (2 km wide) isthmus with a level habitat which is at an elevation of *c*. 5 m a.s.l. and which has a substrate of loam and a few rocks and pebbles; up to 30% of the cover consists of a black and grey crust, which occupies a portion of the polygon surface. Along the fissures outlining the polygons there is a network of herb–moss–lichen sward, in the composition of which *Hylocomium alaskanum, Ditrichum flexicaule* and other mosses participate here and there; on wet sites a mat of fruticose lichens grows in which *Cetraria cucullata* is the most abundant but in which *C. delisei* and *C. islandica* are also common. The cover of herbs amounts to 20%; among these, *Alopecurus alpinus* and *Papaver polare* are the most common, but *Cerastium regelii, Deschampsia caespitosa* ssp. *glauca, Phippsia algida, Poa alpigena, Ranunculus sulphureus, Saxifraga cernua, S. oppositifolia, Stellaria edwardsii* and *Salix reptans* are also encountered. Besides similar associations, occupying plakor habitats, others are met with on the Parizhskaya Kommuna peninsula on wet ground with polygons where, locally, mosses predominate as well as, in particular, a 'black film'. Including the latter, the cover amounts basically to 65–70% but, locally, can be as much as 90–95%. *Deschampsia caespitosa* ssp. *glauca* is the most abundant of the flowering plants here and forms hummocks. The 'black film' almost completely covers the lower parts of slopes, where the snow persists for a long time. On south-facing slopes that are more satisfactorily heated tundra associations occur in which *Salix polaris* appears (Korotkevich, 1958).

The vegetation on the Zhiloy peninsula, which is situated on the northwestern coast of Ostrov Oktyabr'skoy Revolyutsii somewhat south of the Parizhskaya Kommuna peninsula, can also be referred to as 'arctic semidesert'. Safronova (1976) described the development on quaternary sediment here of a patchy grass (*Poa abbreviata, P. alpigena, Alopecurus alpinus*) – herb (*Papaver polare, Saxifraga oppositifolia, S. caespitosa, Cerastium beeringianum* ssp. *bialynickii, Draba oblongata, D. subcapitata* and *D. macrocarpa*) – lichen–moss association with a general cover of from 50 to 85%, of which the mosses represent 35–60%, the lichens 10–40% and the herbs 12–50% of the area occupied by the association. Small hummocks of grass (*Alopecurus alpinus*, etc.) and moss clumps (estimated cover of latter, 50%) with (*Alopecurus alpinus*) – lichen swards between them are characteristic of hollows; locally, *Nostoc commune* can be found. According to Safronova (1976) and Khodacheck (1980), associations with *Salix polaris*, often in considerable quantities, can also be found on the Zhiloy peninsula.

The rich vegetation on the north-western extremity of Ostrov Oktya-br'skoy Revolyutsii is especially striking because Ostrova Sedova lie not far away to the west of the Parizhskaya Kommuna and the Zhiloy peninsulas. The vegetation on Ostrova Sedova is so impoverished that it belongs to the northern belt of the polar deserts (Korotkevich, 1958; Khodacheck, 1980). The vegetation on the Parizhskaya Kommuna penin-sula is also considerably richer than that of the plant communities described by Korotkevich (1958) from the wide coastal plain on the south-eastern part of the island situated further south than this peninsula. In part the difference in soil formation may be of importance. On the Parizhskaya Kommuna and the Zhiloy peninsulas the vegetation has developed on quaternary sediments. According to information published by Korotkevich on the Parizhskaya Kommuna peninsula the sediments take the form of heavy loams with inclusions of boulders and pebbles (occasionally very large boulders are met with) but on the south-eastern plains, as stated by the same author, the level surface of terraces at an elevation of *c.* 25 m a.s.l. are built up of brownish loam with a large amount of calcareous rocks, the presence of which may be unfavourable for the development of a vegetation, especially of mosses. However, such a composition of the substrate does not seem to interfere with the formation of well-developed phytocoenoses of tundra type, including *Salix polaris*, on slopes in the valley of the Ozernaya river. Although Ostrova Sedova are built up of paleozoic, sedimentary rocks containing limestone, there are substrates with an important content of loam. In some habitats which are not very low-lying, loamy substrates are also met with (Korotkevich, 1958). The plant associations developed on these soils are, however, extremely poor. Although we may assume that the composition of the substrate is more favourable for the development of plants on the Parizhskaya Kommuna and the Zhiloy peninsulas in comparison with that on Ostrova Sedova, this may not explain the very sharp difference in degree of richness of the vegetation. An observation by Khodacheck (1980) drew my attention to the interesting existence on these peninsulas of a temperature anomaly, which may explain the increased richness of the vegetation in a more satisfactory manner. This author, who during 1979 and 1980 was stationed on the Zhiloy peninsula in order to investigate its vegetation, made systematic meteorological observations and, according to his data, the temperature on this peninsula during the growing period was almost 5°C higher than on Ostrova Sedova.

This temperature anomaly resulting in a considerably richer vegetation on the Parizhskaya and Zhiloy peninsulas is associated with the presence there of adjacent high (*c.* 600 m) mountains and the inter-montane valley of the Peschanaya. Such a situation frequently promotes the periodic appearance of foehn winds (similar to those documented for the Isefjord on Spitsbergen, for northern Greenland, for islands in the Canadian arctic archipelago and for the valleys between the mountains on Ostrov Vrangelya (see Svatkov, 1970)). In his paper published in 1958 Korotkevich does not mention anything about the presence of foehn winds on Severnaya Zemlya, but in his book published later he writes: 'A particularly important effect results from the foehn winds in the "dry valleys" among the mountains on Victoria Land in the Antarctic as well as those in northern Greenland, on the Queen Elizabeth Islands and on Severnaya Zemlya' (Korotkevich 1972, p. 123). If the courses of the weather components are plotted in respect to the foehn winds (Fig. 42), it can be seen that they cause air temperature to increase, relative humidity to decrease, atmospheric pressure to increase and hours when the sky is clear to become more numerous.

The extrazonal vegetation described by Korotkevich (1958), from a river valley situated on the central plateau of Ostrov Oktyabr'skoy Revolyutsii is still richer. The great depth cut by the valley (Korotkevich, 1958) contributes also to the development of a foehn situation. As a result, plant associations belonging to a type of arctic tundra association can be encountered in such a valley. Thus, in the valley of the Ushakov (a valley deeply cut into the surrounding plateau) a 'spotted tundra' with a general cover of *c*. 80% on a terrace rising to an elevation of 50–60 m above sea level has been described. There was an abundance of mosses, especially of *Tomenthypnum nitens* and *Distichium capillaceum*; fruticose lichens were found in very small quantities (only locally white strips of *Thamnolia* were noticed). Herbs and *Salix polaris* were very well developed. During the daytime a rich flowering of many species was seen: the terrace was outlined by a colourful cover of flowering cushions of *Eritrichum villosum, Novosieversia glacialis, Saxifraga oppositifolia, Artemisia borealis* and *Draba macrocarpa*, etc. In total 27 species of angiosperms have been recorded there; *Salix polaris* was met with in abundance. Unfortunately, nothing is mentioned about the cover of this species, since in the table given by Korotkevich (1958, p. 649) the cover (12%) represents the sum total of that of the *Salix polaris* and *Saxifraga oppositifolia* group because, as mentioned above, Korotkevich erroneously treated the latter species as dwarf shrubs. 'Bog-like' habitats were also encountered in this river valley, which, according to their species composition, do not belong to the region of polar deserts but to the subzone of the arctic tundra. In one of these 'bogs' Korotkevich was able to locate such rarities for the island of Oktyabr'skoy Revolyutsii as *Dupontia fisheri* and *Chrysosplenium alternifolium*.

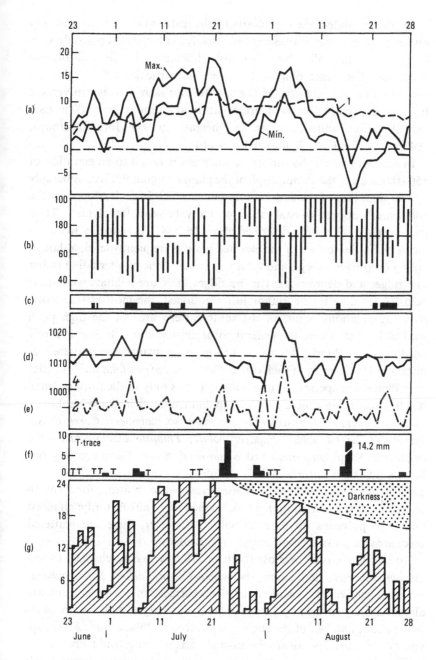

Fig. 42. Graphs representing weather components in relation to foehn winds on Axel Heiberg Island (after Müller, according to Korotkevich, 1972, p. 122). (*a*), air temperature (°C); (*b*), relative humidity of the air (%); (*c*), foehn winds; (*d*), air pressure (mbar); (*e*), wind force (m/s); (*f*), total precipitation (mm); (*g*), duration of clear skies (h); *1*, water temperature of lakes.

As already stated, the complexity of the distribution of the vegetation on the southern islands of Severnaya Zemlya is connected not only with temperature anomalies but also with differences in the soil-forming processes. The vegetation is, however, in particular affected by the formation of *vertical belts*. On Ostrov Oktyabr'skoy Revolyutsii vertical belts have developed on the interior plateau at altitudes up to 300 m a.s.l. and at various altitudes above that height (e.g., on Gora Bazarnaya, 594 m a.s.l. and Gora Byk at 703 m a.s.l.).

In localities where the surface of the plain is raised to an elevation of 150–180 m a.s.l., the composition of the plant communities is considerably poorer than that of the zonal communities which have developed on the wide south-eastern coastal plain of Oktyabr'skoy Revolyutsii. Their characteristics have been described above (p. 134). As stated by Korotkevich (1958), the surface of the elevated plain is frequently level, although cut by deep river valleys, and locally its relief is one of alternating rather low ridges and depressions. On the ridges there are habitats with stone fields. According to the author just mentioned, the most characteristic plant aggregations occur on the south-eastern area of the high plain studied by him, where the general cover amounts to only 1–2%, rarely reaching 3%. Poorly developed low-growing plants are scattered at distances 15–30 cm from each other. *Saxifraga oppositifolia* thrives better here than other species, although it too grows only in the form of small patches measuring a mere 10–15 cm in diameter. Other plants mentioned as growing here are *Cerastium beeringianum* ssp. *bialynickii, C. regelii* ssp. *caespitosa, Draba pohlei, Papaver polare, Phippsia algida, Puccinellia angustata, Saxifraga cernua* and *Stellara edwardsii*. There are very few lichens (mainly *Certraria delisei*) and mosses are extremely rare, a fact which is attributed to the lack of adequate heat or moisture for them. In wetter habitats the vegetation has developed more satisfactorily: on moist loam, on polygons up to *c.* 30 cm in diameter, there are scattered specimens of grasses (*Deschampsia caespitosa* ssp. *glauca* and *Poa alpigena*) and hummocks of moss (built up of *Polytrichum alpestre, Orthothecium chryseum,* etc.); among the lichens *Cetraria delisei, Sphaerophorus globosus,* etc. can be encountered. The surfaces of such polygons are almost entirely covered by a crust of lichens and algae. Adjacent to an ice dome about one half of the surface is covered by 'black film'. *Bryum* sp. and *Polytrichum alpestre* are the most abundant mosses and there are few lichens. Here, among the flowering plants, *Poa alpigena* forms small hummocks and *Cerastium regelii* ssp. *caespitosum* and *Saxifraga cernua* are also met with. However, it was stated that there are few habitats with a better developed vegetation and that, in general, the plant cover was characterized by extreme poverty.

On the even more elevated surface of the periglacial plain (at an altitude of 250–300 m a.s.l.) the vegetation is even more impoverished. Plants are found there only as individual specimens. Even where it is considerably wetter the cover does not reach 10%. Of the flowering plants *Phippsia algida* is the commonest; of the mosses it is *Polytrichum alpinum* and *Ditrichum flexicaule*. In the area of Gora Bazarnaya at an elevation of 350–400 m a.s.l., of the flowering plants only *Saxifraga cernua* can be found and among the mosses only *Rhacomitrium lanuginosum*. Epilithic lichens can be seen on the surfaces of stones (Korotkevich, 1958).

The presence on Ostrov Oktyabr'skoy Revolyutsii of extrazonal habitats with a richer vegetation means that the flora is relatively richer in higher plants than is usual at such a high altitude. As can be seen from Table 6, 75 species and subspecies of flowering plants have been recorded up to the present time from this island. This is a considerable number of taxa for an area situated within the arctic polar deserts. In addition to arctic, high-arctic and alpine–arctic taxa which are absolutely dominant in the composition of the flora, there are also hyparctic–montane species (*Artemisia borealis*) and one arctic–boreal species (*Chrysosplenium alterniflorum*). Both are associated with extrazonal habitats.

The Siberian characteristics of the flora are clearly visible: of the 75 species 53 have a circumpolar distribution area and 22 have a Siberian, a Euro-Siberian, a Eurasiatic or a Siberian-west American (rarely a Eurasiatic-west American) area of distribution (Table 6).

As already discussed the natural conditions on Severnaya Zemlya are extremely variable and full of contrasts. As an example of the contrasts in the natural conditions on Ostrov Oktyabr'skoy Revolyutsii, can be mentioned the fact that as on the western coastal area of the Parizhskaya Kommuna peninsula there is a relatively rich (in relation to such a locality) extrazonal vegetation of the 'arctic semi-desert' type. In even warmer habitats there are tundra phytocoenoses containing *Salix polaris*. There are glaciers which reach out into the sea at roughly the same latitude on the eastern shore of this island. There shelf-ice, formed by floating glacial tongues originating from the Karpinskiy and the Rusanov ice sheets occurs on Zaliv Matusevicha (Semenov, 1970). At the same time, west of and in the immediate vicinity of the Parizhskaya Kommuna peninsula, Ostrova Sedova present an extremely impoverished, limited and poorly developed flora and vegetation.

Just as Zemlya Frantsa-Iosifa has sometimes been called 'the Little Antarctic' (Govorukha, 1970*a*), since a certain similarity can be perceived between it and what is named the 'maritime Antarctic' (i.e., the area including the Antarctic peninsula and a number of islands: South Shetland, South Orkney, etc.), some of the natural contrasts on Severnaya Zemlya bring to mind the similar contrasts found in Greenland.

Mys Chelyuskin

Until quite recently, the vegetation of Mys Chelyuskin was practically unknown and there was only scanty information about it (Kjellman, 1883*a*, 1883*b*; Byalynitskiy-Birulya, 1902; Birulya, 1907). In spite of the opinion of Gorodkov (1935) and Sambuk (1937), who included this area in a subzone of the arctic tundra, I placed it in the polar desert zone when carrying out the zonal division of the Arctic (Aleksandrova, 1971) based on extrapolation, since the data available on its vegetation in relation to the circumpolar-arctic isotherms made me conclude that north of the July 2°C isotherm the tundra landscape must change into a landscape of the polar desert type.

Thanks to the investigations by Matveyeva (Matveyeva and Chernov, 1976; Matveyeva, 1979), the vegetation of the area of Mys Chelyuskin is now one of those best known in the literature concerning the polar desert areas. What is of special importance is the detailed study of the single-tier structure of the plant aggregations and the fact that at five localities where complex permanent investigations were made the total specific composition, including all the cryptogamic components (except for the epilithic, crustose lichens, for which only the general cover was estimated), has been characterized in the same way as that of other plant aggregations found within this area. The flowering plants of the area were studied by Safronova (1979); the composition of the mosses and the liverworts was also elucidated by Blagodatskikh *et al.* (1979) and that of the crustose lichens by Piyn (1979). The soils were studied by Chugunova (1979) and microclimatic observations were carried out and studied in connection with an investigation of the vegetation and soil fauna (Matveyeva and Chernov, 1976; Chernov *et al.*, 1979).

There are no glaciers in this area. The mean temperature in July reaches 1.5°C, in August 0.8°C. Like Severnaya Zemlya, the area of Mys Chelyuskin belongs to the Taymyr–Severnaya Zemlya fold area. The relief is formed by low ridges of base rock outcrops (clayey and calcareous siltstone predominate; more rarely there are broken down quartzite veins and outcrops of dolomites) alternating with rolling areas where the base rock is covered by quaternary maritime sediments in the form of moderately, rarely very heavy, loams (Matveyeva and Chernov, 1976).

Just as on Zemlya Aleksandra, the plant aggregations from Mys Chelyuskin that were studied and described by Matveyeva (1979) belong to three categories: nanocomplexes (according to her terminology the majority of these have a 'continuous polygonal network'), semi-aggregations ('aggregations with a strip and cushion-like, open type of scattered sward') and aggregations where the plant communities are composed of individual specimens growing separately with a general cover of less than 1%.

In aggregations with a continuous cover (nanocomplexes) in this area two types of structure can be distinguished, just as on Zemlya Aleksandra. One is a polygonal network; the other, found in habitats completely covered (although in exceptional habitats interrupted by bare ground) by a closed but thin cover, consists mainly of lichens on dry sites with small blocks and of 'black film' on wet sites with loamy soils.

In the group of nanocomplexes, the continuous. polygonal network type of vegetation seems to be the one most characteristic for Mys Chelyuskin (Matveyeva, 1979). Such a structure is especially typical also of those aggregations which, according to the description given by Matveyeva, are represented by the zonal vegetation within this area.

Fig. 43. Horizontal composition of a moss–lichen aggregation on Mys Chelyuskin containing predominantly *Aulacomnium turgidum*. Dimensions of the plot, 1 × 2 m (according to Matveyeva, 1979, p. 25). *1*, bare ground; *2, Aulacomnium turgidum*; *3, Rhacomitrium lanuginosum*; *4, Polytrichum alpinum*; *5, Tomenthypnum nitens*; *6, Psoroma hypnorum*; *7, Cerastium regelii*; *8, Deschampsia caespitosa* ssp. *glauca*; *9, Stellaria edwardsii*; *10*, fissures, filled with rocks.

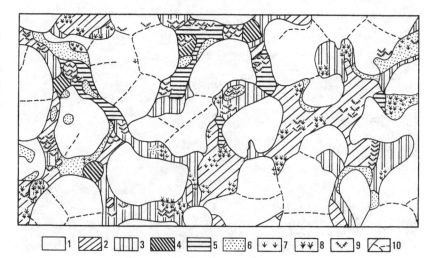

Zonal vegetation

The zonal vegetation is here represented by aggregations developed on a loamy substrate under conditions of moderate humidity (Matveyeva, 1979). One of these has been described in detail (Matveyeva, 1979, pp. 8–9). Loams with a small admixture of rocks have developed fissures around polygons, measuring 50–60 cm in diameter (Fig. 43). Lichen–moss turf appears along the fissures in the form of rather broad strips (Fig. 44), with a general cover of plants amounting to 30–40%. Among the mosses, *Aulacomnium turgidum* is dominant, covering 15% of the site described; scattered patches of *Rhacomitrium lanuginosum* were also noticed. Among the lichens particularly large quantities of *Thamnolia subuliformis, Stereocaulon rivulorum, Psoroma hypnorum, Pannaria pezizoides* are found and *Cetraria delisei*, etc., is also present. *Deschampsia caespitosa* ssp. *glauca, Stellaria edwardsii* and *Phippsia algida* are most abundant among the flowering plants. *Cerastium regelii, Draba oblongata* and *Saxifraga foliolosa* are also met with.

It is interesting to compare the zonal plant aggregations of the Mys Chelyuskin area with the zonal vegetation on Zemlya Aleksandra (see p. 77). In spite of the facts that both belong to nanocomplexes with a continuous polygonal-network vegetation sward, that mosses and lichens seem in both cases to be the basic components of the plant aggregations

Fig. 44. A moss–lichen aggregation containing predominantly *Aulacomnium turgidum* on Mys Chelyuskin (photo.: N. V. Matveyeva).

and that the flowering plants are locally insignificant, there are differences between them. These result from the fact that Zemlya Aleksandra is situated within the northern belt of the arctic polar deserts while Mys Chelyuskin lies within the southern belt and that these areas belong to different geobotanical provinces of the polar deserts: in the former case to the Barents province, in the latter case to the Siberian province.

The more southerly position of Mys Chelyuskin means that there the strips of moss–lichen sward are broader and occupy a larger surface area (compare Figs. 29 and 43). It is important that *Aulacomnium turgidum* has increased in abundance (it belongs to the *Hylocomium alaskanum* eco-group; see p. 46); it is met with only as single individuals in the zonal associations on Zemlya Aleksandra, where *Ditrichum flexicaule* plays the major role. In addition it is noteworthy that among the flowering plants in the aggregations on Mys Chelyuskin the primary position is taken by *Deschampsia caespitosa* ssp. *glauca*, while on Zemlya Aleksandra *Deschampsia caespitosa* is not a member of the zonal aggregations and seems in general to be an extremely rare plant because the climatic conditions are too severe for it.

A still more significant difference in the composition of the plant aggregations is connected with the fact that Mys Chelyuskin and Zemlya Aleksandra belong to different geobotanical provinces. The more humid climate of Zemlya Aleksandra (i.e., a significant number of days with fog and rime, a predominance of low clouds, frequent precipitation, a high relative humidity) promotes a better development on the polygons of a crust of crustose lichens together with liverworts, while on Mys Chelyuskin the polygons are to a certain extent quite bare. The essential difference lies in the specific composition of the surface crusts: not a single species of *Pertusaria* is mentioned from the plant aggregations on Mys Chelyuskin, but *Ochrolechia frigida* is marked as present (cover: < 1%), while on Zemlya Aleksandra these species form the basic, brightly coloured 'background crust', the cover of which amounts to 45% in the zonal aggregations.

The amphiatlantic species *Cerastium arcticum*, which is not found on Mys Chelyuskin, is a member of the zonal plant aggregations on Zemlya Aleksandra. The composition of the rest of the flowering plants (with the exception of *Deschampsia caespitosa* ssp. *glauca*) is in general just like the composition of the crustose lichens. The large amounts of *Thamnolia* do, however, attract attention. This genus is rarely met with on Zemlya Aleksandra and is there represented only by *Thamnolia vermicularis*, while on Mys Chelyuskin it is widely distributed in the form of *Thamnolia subuliformis*. The abundance of *Aulacomnium turgidum* lends a 'Siberian' colour' to the zonal aggregations on Mys Chelyuskin, since in general it plays a greater role in Siberian associations within the Arctic than it does in Europe.

As mentioned above (p. 49), the major part of the ecological groups distinguished by myself on Zemlya Aleksandra can be observed also among the plant aggregations on Mys Chelyuskin, although there are differences in the composition of the eco-groups due to the fact that the two areas belong to different provinces of the polar deserts. Since a relatively large number of geobotanical data have been collected from the Mys Chelyuskin area, including complete lists of species, a classification can be made on the same basis as that set up for Zemlya Aleksandra. As a result it becomes thereby necessary to distinguish both the groups belonging to the type of nanocomplexes and those belonging to the type of open aggregations into taxonomic units of higher rank.

Nanocomplexes which display a polygonal network structure, but which are more satisfactorily developed than the zonal aggregations, were described by Matveyeva (1979, pp. 8–9, plot no. 2) from a level surface on a dolomite ridge on Mys Chelyuskin (according to Matveyeva they also include aggregations that are rather rare among those observed in this locality). There the substrate consists of rocks mixed with fine soil

Fig. 45. Lichen–moss aggregation with a polygonal network type of vegetation on a dolomite outcrop in the area of Mys Chelyuskin (photo.: N. V. Matveyeva).

fissured into rather small polygons (Fig. 45) so that the larger blocks are situated along the fissures and the fine soil forms the polygons. The extensive development of mosses (cover: *c.* 50%) is typical. *Ditrichum flexicaule* and *Distichium capillaceum* predominate. The lichens are less abundant, covering about 10%. Those dominating are *Cetraria delisei, C. islandica* var. *polaris, Stereocaulon rivulorum* and *Thamnolia subuliformis.* Among the flowering plants, represented by about ten species, *Saxifraga oppositifolia* (cover: 3%) is the commonest; the rest of these species (*Alopecurus alpinus, Deschampsia caespitosa* ssp. *glauca, Phippsia algida, Stellaria edwardsii, Cerastium regelii, Papaver polare, Draba oblongata, D. subcapitata, Saxifraga nivalis* and *S. cernua*) have a cover of less than 1%. These aggregations differ essentially from tundra associations because of a constantly lesser participation of angiosperms (which, besides, are much smaller and do not display any tendencies to form even fragments of independent tiers) and because of the complete absence of any arctic–alpine dwarf bushes.

Matveyeva (1979) also described nanocomplexes with a polygonal network form, but with less well-developed strips of sward constituting the network. To such nanocomplexes frequently belong plant aggregations that are met with on loamy ground: here *Orthothecium chryseum* dominates and the width of the mossy strips is 10–20 cm, while the general cover amounts to 30–40%. On habitats with siltstone outcrops and patterned ground with rocky substrates, where due to cryogenic processes the larger rocks have been sorted out from the fine soil, plants cover 30–40% of the surface. Lichens predominate and individual angiosperms occur. On stony, polygonal ground, built up of dolomite, there are strips of vegetation in relatively snow-rich habitats, where dark-coloured species of *Cetraria* predominate, while the patches of fine soil are covered by a crust of *Toninia lobulata;* in such cases the plants cover 20–30% of the surface.

A second type of closed cover (a continuous one, not a polygonal network type) is, according to the descriptions by Matveyeva (1979), met with on siltstone outcrops in habitats with a level surface consisting of small blocks, 3–4 cm across, and rarely of larger ones (30–40 cm in diameter). There the cover of plants amounts to 80–90%. (A few bare rocks can occur within the closed cover.) Lichens predominate, consisting of the majority of the species mentioned above with *Cetraria* spp. being the commonest; mosses, among which the epilithic *Andreaea rupestris* is encountered, are less abundant, and so are the flowering plants. Such an aggregation was described in detail by Matveyeva (1979, pp. 7–10, plot no. 5).

Other observations by Matveyeva (1979) concerning the aggregations include the category of semi-aggregations, where there is no continuous cover but a type of sward distributed in the form of strips and cushions, i.e., individual strips of flowering plants and cushions of mosses and lichens sporadically scattered over the surface. Her plots nos. 3 (Fig. 46) and 4 (Fig. 47) contain such semi-aggregations. The vegetation that develops on polygonally fissured loam in habitats where the snow persists longer belongs to the category of very impoverished semi-aggregations (Fig. 48).

Fig. 46. Lichen–moss polar desert vegetation, forming strips and cushions (a semi-aggregation) in the area of Mys Chelyuskin. Dimension of the plot, 5 × 5 m (according to Matveyeva, 1979). *1*, bare ground; *2*, mosses (*Rhacomitrium lanuginosum, Aulacomnium turgidum*, etc.); *3*, lichens (*Stereocaulon rivulorum, Thamnolia subuliformis*, etc.); *4*, *Cerastium regelii, Stellaria edwardsii, Phippsia algida; 5*, fissures, filled with rocks; *6*, secondary fissures due to drought.

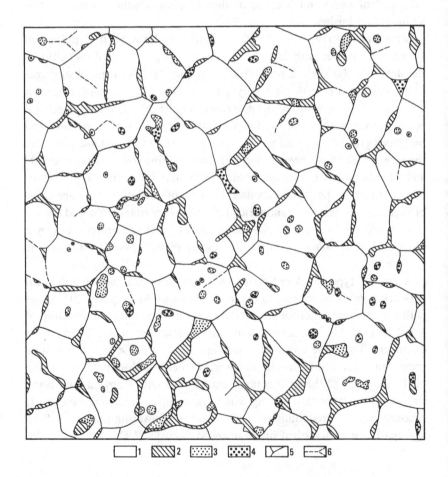

On a small portion of Mys Chelyuskin, within a zone of eroded siltstones containing ferro-carbonates, the plant cover is rather poorly developed and consists of aggregations containing only a few species of lichens and mosses; flowering plants are absent there (Matveyeva, 1979). In habitats where water remains standing, weakly developed fragments of mossy small bogs are met with; there are no flowering plants (Matveyeva, 1979, pp. 22–23).

Fig. 47. Moss–lichen aggregation with vegetation distributed as strips and cushions (a semi-aggregation) in the Mys Chelyuskin area. Dimension of the plot, 5 × 5 m (according to Matveyeva, 1979). *1*, bare, loamy ground; *2*, mosses (*Ditrichum flexicaule, Orthothecium chryseum, Bryum tortifolium*, etc.); *3*, lichens (*Stereocaulon rivulorum, Cladonia pocillum, Thamnolia subuliformis, Toninia lobulata*, etc.); *4*, flowering plants (*Stellaria edwardsii, Cerastium regelii, Phippsia algida*, etc.); *5*, fissures in the ground.

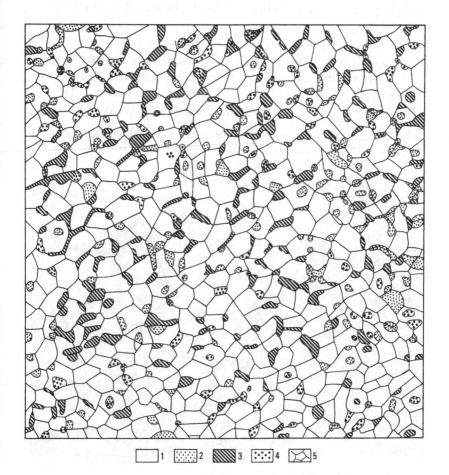

As far as arctic–alpine dwarf shrubs are concerned (i.e., the characteristic components of arctic tundra phytocoenoses), they are not present in these locally very well studied plant associations; nor do they feature in the composition of other aggregations described and relatively well studied by Matveyeva (1979). In the conspectus of the flora published by Safronova (1979), *Salix polaris* and *Dryas punctata* are mentioned among a number of arctic–alpine dwarf shrubs. Matveyeva states, however, that in some localities habitats are found 'with a comparatively continuous cover in respect to the amount of the moss sward developed, but differing from the tundra associations particularly in respect to the dwarf shrubs *Salix polaris* and *Dryas punctata*, which are so typical of the subzone of the arctic tundra but here occur in extremely low numbers only' (Matveyeva, 1979, p. 27). Concerning the presence in this area of grass–willow–moss–lichen tundra associations with *Salix polaris*, these are apparently met with only in local habitats that are more effectively heated (Safronova, 1979, p. 51).

The flowering plant flora of Mys Chelyuskin has, just like that of Severnaya Zemlya, a clearly visible Siberian character: there are 13 species with Siberian, Siberian-American and Eurasiatic-Siberian areas of distribution. The remaining 44 species are circumpolar (Table 6). There are considerably fewer species (57) than on Ostrov Oktyabr'skoy Revolyutsii

Fig. 48. Very impoverished semi-aggregation on loam fissured into polygons in a habitat where the snow lasts for a long time, in the Mys Chelyuskin area (photo.: N. V. Matveyeva).

in Severnaya Zemlya (seventy-five sp. and ssp.). The following angiosperms, mentioned as being present on Ostrov Oktyabr'skoy Revolyutsii,
are lacking on Mys Chelyuskin: *Deschampsia brevifolia, Festuca hyperborea, Poa lindebergii, Pleuropogon sabinii, Puccinellia phryganodes, P.
angustata, Carex ensifolia* ssp. *arctisibirica, Minuartia rubella, Stellaria
humifusa, S. laeta, Eutrema edwardsii, Braya purpurascens, Draba pilosa,
D. barbata, D. lactea, D. pseudopilosa, D. pohlei, D. kjellmannii, Parrya
nudicaulis, Saxifraga rivularis, Chrysosplenium alternifolium, Potentilla
pulchella* and *Artemisia borealis*. The following species are met with on
Mys Chelyuskin but not on Severnaya Zemlya: *Carex stans, Eriophorum
scheuchzeri, Polgonum viviparum, Minuartia macrocarpa, Stellaria ciliatosepala, Caltha arctica, Myosotis asiatica, Saussurea tilesii* and *Senecio
atropurpureus*.

Of the differences enumerated only one is remarkable: in the area of
Mys Chelyuskin the flora is purely arctic, consisting of arctic, high-arctic
and arctic–alpine species. The two species which have a more southerly
type of distribution, i.e., *Chrysoplenium alternifolium*, an arctic–boreal
species, and *Artemisia borealis*, a hyparctic-montane species, are not met
with on Mys Chelyuskin although they occur on Ostrov Oktyabr'skoy
Revolyutsii because of the existence there of habitats more favourable to
them and with an extrazonal vegetation. The fact that a large number of
species is found on Ostrov Oktyabr'skoy Revolyutsii can, of course, be
supported by two simple explanations: firstly, the area of this island is
larger than that occupied by Mys Chelyuskin; secondly, there is a greater
variation of habitats on the island.

Ostrova De Longa

Five small islands belong to the Ostrova De Longa, i.e., Ostrova
Genrietty, Zhannetty, Bennetta, Zhokova and Vil'kitskogo. These are
included in the Novosibirskiye Ostrovo under the epithet 'Malye Ostrovo'
(the small islands). On the first three, i.e., those that are most northerly, ice
cover has developed (see above, p. 8); Ostrova Zhokova and Vil'kitskogo
are free of ice. Concerning the vegetation on these islands, data have been
published by Kartushin (1963*b*) on the vegetation of Ostrov Bennetta and
some brief information has been given by Korotkevich (1958; 1972) by
Semenov (1968) and by Sisko (1970).

Kartushin (1963*a*), who studied the glacier on Ostrov Bennetta during
1956, briefly described the vegetation of this island and collected a
herbarium which was worked over by Tolmachev (1959). However, the
characteristics of the vegetation, as reported by Kartushin, are contro-
versial. On the one hand Kartushin states: 'Ostrov Bennetta is character-

ized by an extremely scattered plant cover, composed of crustose and
other lichens, relatively many species of mosses but very few flowering
plants' (Kartushin, 1963a, p. 178). On the other hand, he mentions the
presence on the island of 'a polygonal herb–dwarf shrub–moss tundra',
where the dominant role is played by mosses, 'in the protection of which'
grow *Alopecurus alpinus, Salix polaris, Ranunculus sulphureus, Saxifraga
caespitosa, S. cernua, S. foliolosa, Poa arctica* and some lichens, including
Thamnolia and *Dactylina* ssp. In addition this author states that in the
valley of the Lagernaya he saw a *Dryas* tundra; here 50% of the cover
consisted of *D. octopetala* s. lat., mosses played an important role and
Salix polaris, Alopecurus alpinus, Poa arctica, Potentilla hyparctica and
Papaver polare participated in additon to a number of lichens. Further-
more, he also discovered, in the valley of the Trollevskaya, some small
(20–30 cm deep) ponds, the shores of which were covered by a dense
growth of *Alopecurus alpinus* reaching 20 cm in height. This vegetation
was evidently of tundra type. The question arises: 'On the basis of this
information, should Ostrov Bennetta actually belong to the tundra zone?'

The greatest botanical authorities who have studied the high-latitude
Arctic, i.e., Korotkevich (1958, 1972) and Semenov (1968), provided
additional information and placed the Ostrova De Longa in the zone of
the arctic polar deserts. Korotkevich (1972), in particular, included them
among 'the most typical' areas belonging to the moss–lichen subzone of
the polar deserts. For this he did not use as wide a concept of the arctic
polar desert type of vegetation as that of Gorodkov (1958) but rather
based his classification on principles conforming with my own interpre-
tation (see Korotkevich, 1958). Holding a similar position, Semenov
(1968) placed the Ostrova De Longa in the southern belt of the arctic
polar deserts. Sisko (1970, p. 443) stated: 'On the Ostrova De Longa there
is no continuous vegetation cover. The plants are distributed in the form
of strips along fissures, and single plants and small groups thereof are
scattered on the bare, clayey-stone soil.'

In order to explain these contradictions, it is necessary to consider the
opinion proposed by Kartushin, i.e., that tundra type phytocoenoses are
present in the river valleys of Ostrov Bennetta. If the fact that the ice sheet
on this island reaches an altitude of 426 m a.s.l. is taken into considera-
tion, then it could indeed happen that in deeply cut valleys foehn winds
periodically develop (see pp. 11–12), resulting in a total amount of heat
that is locally favourable for the development of an extrazonal vegetation
of tundra type. Such an explanation is logical.

The existence of extrazonal habitats which are considerably warmer
than the surrounding landscape is confirmed by the presence in the flora

Fig. 49. Area of distribution of *Ranunculus sabinii* R. Br. (according to Hultén, 1968; *Arctic Flora of the USSR*, vol. 6, 1971; Polozova and Tikhomirov, 1971; Tolmachev and Shukhtina, 1974).

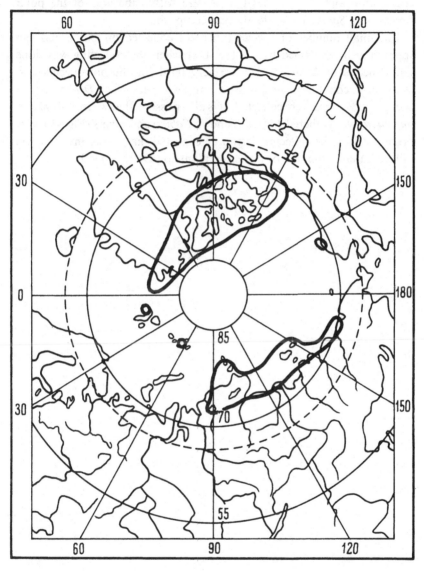

on Ostrov Bennetta of such a southerly (in relation to the arctic polar deserts) species as *Nardosmia frigida*, which has an arctic–boreal area of distribution and of a species that is rare within the area of the polar deserts, i.e., *Saussurea tilesii*, another composite.

The small number of species (only 20) collected on this island can apparently be explained by the fact that when the collecting was done some species of *Draba*, which are very hard for the non-specialist to distinguish, could have been missed: only two species, i.e., *Draba barbata* and *D. macrocarpa* were collected. It is also possible that when *Ranunculus sulphureus* was gathered *R. nivalis* and *R. Sabinii* and others (Fig. 49) were overlooked by the collector. It can be expected that future investigators will find a large number of flowering plants on Ostrova De Longa.

5

Seasonal development of the flowering plants within the polar deserts

Hardly any information has been published concerning the seasonal development of the plants in the polar deserts. Only the role of 'greenhouses' of snow and ice during the springtime for the generation of plants has been commented upon (Leonov, 1953; Govorukha, 1970*a*) and there have been a few reports giving the dates when some species were found flowering or setting seed. For example, Tolmachev and Shukhtina (1974) mention the dates of collection of plants at the flowering stage on Ostrov Kheysa (in the archipelago of Zemlya Frantsa Iosifa), including those for *Saxifraga platysepala* (12 August), *S. caespitosa* ssp. *exaratoides* (19 July and 27 August) and *Ranunculus sabinii* (12 August); one of the specimens of the latter, collected on the same date, was at the fruiting stage.

The study of the seasonal development of plants at such high latitudes and under such extremely unfavourable conditions is, nevertheless, of exceptional interest because it can help to explain the adaptability of plants to conditions of an almost insufficient amount of heat.

The study of the seasonal development of the plants was conducted by myself on the island of Zemlya Aleksandra. These data have not been published previously. They cover the period from 31 May to 3 September 1959. Besides information on the phenological state of the plants, which was obtained during investigatory excursions and when doing geobotanical relevés, the material in question furnished information about a number of individual plants, series of which were identified by consecutively numbered tags.

As in other areas of the Arctic, the first signs of the reawakening of the vegetation were observed, as also mentioned by Leonov (1953) and Govorukha (1970*a*), in the 'greenhouses' and 'caves' of ice briefly described above (p. 24). Further exploration revealed extremely interesting phenomena in respect to the acclimatization of the plants in the course of their seasonal development to conditions during the short and very chilly vegetative period characterized by stormy winds accompanying the always unstable weather, with frequently occurring frosts and snowfalls.

Seasonal development of the flowering plants on the island of Zemlya Aleksandra

Gramineae

Deschampsia caespitosa (L.) Beauv. ssp. *glauca* (Hartm.) Hartm. is a densely caespitose, almost circumpolar arctic–alpine species. It was found a few times, always in a sterile condition. Any signs of regenerating shoots from previous years were lacking. No further observations were made.

Poa alpigena (Fries) Lindm. is an arctic–alpine species with long rhizomes which was found growing here and there in small stands. It was met with very rarely and only in conditions of the most favourable soil types or where the snow persisted until 10 or 15 June. It grows in the form of loosely continuous small patches (small stands), to a height of 5–6 cm and produces from one to eight stolons. The individual stands are joined by slender rhizomes, from 2 to 12 cm long.

It had survived the winter with the very youngest leaves on some of the shoots in a green condition (the length of the portion of green leaf buried in the snow amounted to 12 mm, while the entire leaf reached a length of 30 mm). On 12 June when specimens of *P. alpigena* were being examined, a fresh green leaf, 5 mm long, was observed: on 15 June it was 7 mm long, and by 19 June it had grown to 20 mm; the tip of the leaf was tinted with anthocyanin. By 1 July the same specimen had sprouted another new leaf, *c.* 7 mm long. By 9 July the fresh leaf which had appeared first had grown to 32 mm, the other one to 28 mm. The overwintering, green basal leaf was drying up: its green portion had shortened to 3 mm. On 9 June small groups of new shoots emanating from the nodes had been observed, the length of which on that date was from 0.7 to 1.5 cm. On 23 June the green portion of the surviving basal leaf on the specimen of *P. alpigena* under observation had completely dried up. The shoots had three new leaves: a lower one (appearing first) 33 mm long, a second one 38 mm long and a third (the youngest one) 3 mm long. A short regenerating shoot had appeared, in the form of a culm. This culm did not exceed 8 mm but remained at the stage of a small culm; it never produced a panicle.

In some other habitats where *P. alpigena* occurred, spikelets were also not observed, but in one case a remnant of a dead, regenerative shoot from a previous year was discovered. It measured 40 mm long. It had a

panicle without seeds and had evidently survived from an earlier year
under the snow in the panicule stage.

Poa abbreviata R. Br. is a compact caespitose, arctic–alpine species
which is especially characteristic of high-arctic areas (Fig. 50). It can be
characterized as moderately chionophobic and is found on Zemlya
Aleksandra almost exclusively in relatively dry habitats with little snow,

Fig. 50. Area of distribution of *Poa abbreviata* R. Br. (according to *Arctic flora
of the USSR*, vol. 2, 1964; Hultén, 1968; Polozova and Tikhomirov, 1971).

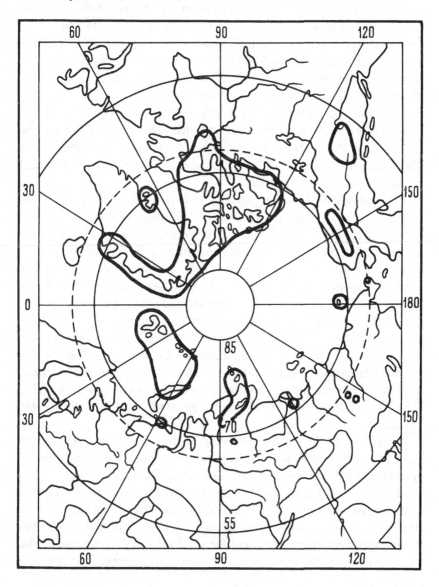

or where the snow depth does not exceed 25 cm and where the snow melts away between the end of May and 22 June.

P. abbreviata had overwintered with the lower quarter to one third parts of the youngest leafy shoots still green. The length of the surviving green leaves was *c.* 12 mm. It appeared very early from below the snow where this started to melt at the end of May. On 2 June there were already fresh leaves, *c.* 2 mm long, on plants in a patch thawed out near Zaliv Dezhneva. The substrate on which the stands of *P. abbreviata* were found was on that day thawed to a depth of 5–6 cm. By 12 June the new leaves on these tufts had grown to 10 mm, by 15 June to 16–17 cm. On 19 June the length of the fresh leaves was 18–22 mm, and on 6 July it was 22 mm. On 6 July the beginnings of small panicles were discovered, on 14 July there were many panicles and on 23 July the flowering started. On 7 August the flowering stage was finished (the panicles closed up).

Under such conditions – apparently those most favourable for the development of *P. abbreviata* – mature little seeds were formed but the dispersal of the seeds seemed, in general, to occur only during the following year. Under other, less favourable conditions, the phenological stages followed at a slower pace.

It is noteworthy that the greening and the subsequent sprouting of culms and the development of panicles take place at first in the lower portion of the tuft, which lies close to the ground. Thus, on 23 July, in one of the stands of *P. abbreviata* studied the first flowers of the panicle opened up and it was possible to observe that the panicle had developed its open flowers and visible anthers only in the lower portion of the tuft and on horizontal growth pressed to the ground and pointing towards a row of stones. Levkovskiy *et al.* (1981) discovered what appeared to be facultative cleistogamy in *P. abbreviata* on Ostrov Vrangelya. Apparently, in the case described there were also on Zemlya Aleksandra unopened cleistogamous flowers in the upper portions of the tufts, which receive less heat, alongside open chasmogamous flowers in the lower portions of the tufts. This occurred also in other stands, growing under less favourable conditions.

In addition to the glaucescent living leaves of the current year, in the tufts of *P. abbreviata* studied there were many dead leaves from previous years (from 2 to 3 cm long): these assured a more satisfactory preservation of the heat, both that absorbed by the stand from solar radiation and that radiating from the ground. Another favourable effect of the mass of dead leaves was also discovered: on 26 and 27 June there was a snowfall accompanied by strong winds and the small tufts of *P. abbreviata* on the slope of Glavnyy Val were covered by snow. But the leaves from previous years, protruding on different sides and keeping the snow off, made the development of a thin, frozen crust around the stand possible. Under this crust there was a cavity, in which the basal portion of the tuft was hidden; the tuft thus formed its own heated shelter. On 28 June the snow was shaken off the upper portion of the stand and on 29 June the entire tuft was free from snow. It was then possible to see many greenish glaucescent leaves and regenerating shoots on its lower portion growing out from the primordial stage.

Phippsia algida (Soland.) R. Br. is a densely caespitose, circumpolar–arctic species (Fig. 20). It is especially characteristic of high-arctic areas and is the most widely distributed angiosperm on Zemlya Aleksandra. It usually avoids dry habitats with little snow, but individual specimens can sometimes be found in such habitats. In the patches which thawed out the earliest it started to appear only around 10 June, i.e., later than most other species. It was most abundant in habitats where the snow which had fallen over the winter was from 30 to 60 cm deep and where it disappeared within the period 1–15 June; it was also present, in lesser amounts, in habitats where the snow was up to 1 m deep and had melted only by 20 July but there the tufts of this species were sterile. Where the snow persisted even longer, it was not found.

Thus, under extreme conditions, *P. algida* displays characteristics of chionophily and a striking ability to make use of a brief growing period, 50–70 days or less in length, and with a very low total amount of heat.

Three growth forms of *P. algida* were observed on the island. The first one was the normal morphological development of the tuft with vegetative and regenerative shoots; the diameter of the turf was 5–6 cm, occasionally ranging from 3 to 7 cm, and the length of the leaves amounted to 2–4 cm. This condition was met with in the most favourable habitats, particularly between pebbles on the Glavnyy Val.

The second growth form of *P. algida* was as miniature, round tussocks, 5–15 cm in diameter. This was found in habitats with wet, sandy-pebbly and loamy-clayey depressions (Fig. 32). Such tussocks were best developed where the snow disappeared between 1 and 10 June. They frequently occupied 7–10% of the surface (see p. 110) and many sprouted regenerative shoots. On the oldest tussocks, found in a senescent condition, the centre had died off so that the periphery was particularly exposed and would thus die later on; the dead portion of the tussock was covered by 'black film'. Where the snow melted off around 15 July, tussocks of *P. algida* could be found in analogous habitats, but there they were sterile and covered up to only 5% of the surface. Where the snow disappeared even later, the cover fell to 1% (see p. 113).

The third growth type took the term of individual, small specimens of *Phippsia algida*. This was met with all over the island in very variable habitats. In the majority of the cases these specimens were sterile. They were frequently represented by just a small number of crumpled leaves a few mm long. In a few cases such specimens were met with in relatively great abundance in spite of the fact that they covered only an insignificant portion of the surface (see p. 113).

P. algida survives the winter with the lower one third to one half of the tuft green, the youngest leaves appearing on the shoots, one or two on each. The length of the overwintering portions was 5–10 mm while the total length the leaves reached was, in general, 20 mm. These leaves were hidden among dead leaves from previous years, firmly adhering to the tuft. Their length amounted to *c.* 20 mm.

The regeneration of the vegetative parts started in the spring, first of all inside 'greenhouses' of ice, of which a number of observations bear witness. Thus, e.g., on 10 June, a sterile tussock of *P. algida* was observed; it had just emerged from an 'ice greenhouse' and a portion of it was still below a crust of ice (about 0.5 cm below it). There were no new leaves on the majority of the shoots in this tussock, but among the overwintering leaves the youngest were green along one third of their lower part. Fresh green leaves with a 'pistachio' nuance and with tips tinted with anthocyanine, were observed on only two shoots in a tussock near stones; on one of these the leaves were up to 1.5 mm long. By 12 June their length had increased to 4 mm, by 13 June to 5 mm and by 15 June to 6 mm. By 16 June they were 7 mm long.

On 16 June there were signs of the greening of *P. algida* in habitats free of snow; in spite of the fact that the yellowish colour of the previous year's leaves still predominated, the green leaves among them could already be seen; the length of the fresh leaves amounted in a number of cases to 18–19 mm.

The earliest appearance of a short culm, hidden among the leaves inside the turf, was noted on 22 June on a specimen growing under favourable conditions. At the same time the basal leaves, which had survived the winter in a green state, began to shrivel up. (On the same date the sites with the greatest abundance of *P. algida* were still hidden below the snow.) On 22 July the beginning of a panicle was observed and on 31 July the first flowers appeared, with the anthers showing; on 6 August both anthers and stigmas were visible. The leaves had become reddish, i.e., started to show autumn coloration. On 15 August the spike closed up but the anthers continued to hang on; hidden among the leaves the regenerating shoots were flush with the leaf tips and only in rare cases overtopping them by a few millimetres. Thus, the development of the regenerating shoots for setting seeds proceeded inside the tussock under the protection of the leaves, both living and those dead and dried up from previous years. Later, the panicle, in which, of course, the set seeds had started to grow slightly, began to push up above the leaves of the tussock.

P. algida flowered considerably later than other species (Fig. 51) and became covered by snow when the seeds were still not ripe. The regenerating shoots, which set the seeds, remained alive and survived the winter. After the snow had melted away, it was found that the surviving culm, with the panicle pushing up above the tussock by almost its entire length (0.8–1.2 cm), had also survived, that the upper portion of the sheaths was green and that the crumpled leaves had an anthocyanine coloration. The higher the panicle had pushed up above the tussock, the sooner the shoots of the previous year started to dry out. In most cases the wilting of the shoots carrying the ripe seeds occurred when the new generation of shoots began to develop inside the tussock. The previous year's panicles on yellowed stalks, carrying compactly shaped, tiny seeds, could be found in August.

As mentioned by Middendorf (1867), a number of authors have expressed the opinion that in the tundra seeds ripen below the snow. This has, however, not been verified experimentally. Rather, this opinion, which has been perpetuated through the literature from one author to another, appears to be one of those non-verifiable ideas that do not correspond to reality, i.e., that seeds should ripen only during the following season. The fact that in the tundra seeds can ripen during the second season has been confirmed by Khodachek (1970), who stated: 'Late flowering species do not often succeed to complete all the stages during one season. The ripening of the seeds and their dispersal take place in such cases during the following summer' (Khodachek, 1970, p. 995).

On Ostrov Vrangelya the presence of cleistogamy has been observed (by Levkovskiy *et al.*, 1981) in both *Phippsia algida* and *Poa abbreviata*. It is entirely possible that a part of the *Phippsia algida* tussocks carry cleistogamous flowers also on Zemlya Aleksandra.

When preparing for the winter, the leaves of *P. algida* began to shrivel up by 20 August, starting at the tips, which took on a yellowish colour. One or two of the youngest leaves on each of the shoots stayed green at their bases. Seed production occurs generally in August, mainly one year after the seeds were originally set. The seeds are dispersed by the wind and are able to germinate under the most unfavourable conditions: they produce locally a mass occurrence of sterile crumpled specimens, as mentioned above. Such individuals do not survive for long but are able to

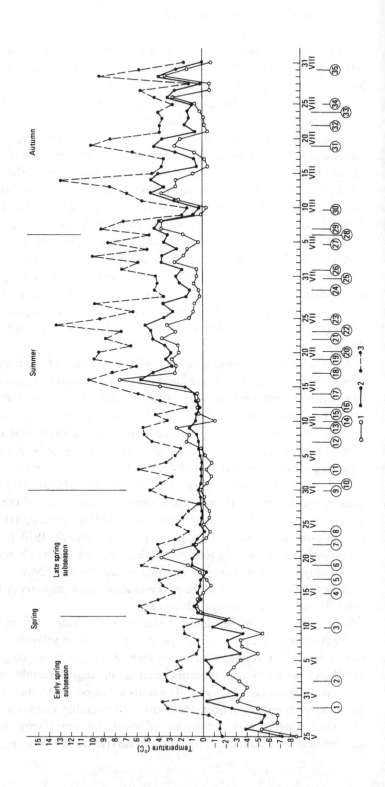

Fig. 51. Seasonal development of flowering plants on Zemlya Aleksandra during 1959. *1*, mean daily air temperature in the meteorological screen; *2*, mean daily temperature at the soil surface; *3*, maximum temperature at the soil surface. Figures below the abscissa indicate days (Arabic) and months (Roman). Figures in circles: *1*, start of vegetative regeneration in patches thawed out and in 'greenhouses' or 'caves' of ice; *2*, new leaves visible on *Poa abbreviata, Phippsia algida, Cerastium arcticum, Minuartia rubella, Stellaria edwardsii, Papaver polare* and *Saxifraga caespitosa* in thawed out patches and leaf buds sprouted on *S. cernua* and *S. hyperborea*; *3*, leafbuds started to open on *Saxifraga cernua* and *S. hyperborea*; *4*, majority of the flowering plants in thawed out patches turned green (new green leaves still did not cover last year's shoots), buds of *Cochlearia groenlandica* started to open, shoots began to show on *Poa abbreviata* and new, vegetative shoots appeared on *Luzula confusa*; *5*, buds on *Draba pauciflora* loosened up; *6*, the first flowers appeared on *Cochlearia groenlandica*; *7*, shoots started to appear on *Phippsia algida*, new vegetative shoots appeared on *Stellaria edwardsii*; *8*, the first flowers of *Draba oblongata* and the loosening of buds on *Draba macrocarpa* were observed; *9*, the first flowers of *Draba macrocarpa* were seen; *10*, the first flowers of *Draba pauciflora* and *D. subcapitata* appeared; *11*, the first flowers of *Saxifraga oppositifolia* were seen; *12*, the first mass appearance of opening buds on last year's stems of *Saxifraga cernua* were noted; *13*, the first flowers of *Saxifraga caespitosa, Cerastium arcticum* and *Luzula confusa* appeared, soft buds were seen on *Saxifraga cernua* and there was a mass appearance of new shoots on small tufts of *Poa alpigena*; *14*, pods started to form on *Draba subcapitata*; *15*, the first flowers of *Cardamine bellidifolia* and *Saxifraga hyperborea* appeared, the first pods formed on *Draba macrocarpa*; *16*, new stalks appeared in tussocks of *Saxifraga cernua*; *17*, the beginning of mass flowering of *Saxifraga caespitosa* and *Cerastium arcticum*, the mass appearance of *Phippsia algida* panicles and the first appearance of pods on *Cochlearia groenlandica* were recorded; *18*, the first flowers of *Saxifraga nivalis* appeared and pods began to form on *Draba paucifolia*; *19*, the first flowers were observed on *Papaver polare*; *20*, pods began to form on *Draba oblongata*; *21*, the mass flowering of *Papaver polare* started; *22*, *Poa abbreviata* flowered, culms of *Poa alpigena* appeared and seed capsules started to form on *Saxifraga caespitosa*; *23*, soft buds appeared on *Stellaria edwardsii*; *24*, the first flowers of *Saxifraga cernua* appeared; *25*, the first flowers of *Phippsia algida* were observed; *26*, capsules started to form on *Saxifraga hyperborea*; *27*, the mass flowering of *Phippsia algida* began and pods began to form on *Cardamine bellidifolia*; *28*, autumnal colours began to appear; *29*, *Saxifraga cernua* started to flower en masse and capsules began to appear on *Luzula confusa*; *30*, the mass flowering of *Papaver polare* ended; *31*, the pods of *Cochlearia groenlandica* started to open; *32*, external leaves on shoots of *Phippsia algida* began to wilt, turning a yellowish colour; *33*, some *Draba macrocarpa* pods began to open; *34*, the unopened buds of *Stellaria edwardsii* dried up, the majority of the pods of *Draba oblongata* and of *D. pauciflora* had opened; *35*, plants were showing signs of the approach of winter: all leaves of *Saxifraga hyperborea* had wilted, become brownish and covered the next year's leaf buds, while on *S. nivalis* the external leaves of rosettes were dark brown and dead, and internal ones (to overwinter alive) had green upper sides with brightly coloured undersides, etc.

produce new shoots inside their sheaths. Therefore it is frequently possible to discover a kind of miniature 'fairy-ring', from 6 to 8 cm in diameter (cf. Fig. 34), where the first shoots formed within these rings and even the secondary shoots have already died off, but where along the periphery of the rings strongly stunted shoots, on each of which there are two to three leaves as fine as hairs and only a few millimetres in length, appear at a greater or lesser distance from each other.

Juncaceae

Luzula confusa Lindb. is a loosely tufted circumpolar, arctic–alpine species. It is very rarely found on Zemlya Aleksandra and then only under the most favourable conditions in habitats with scanty snow. The most highly developed, but rather small, patches of *L. confusa* were encountered along the shore of Zaliv Dezhneva at an elevation of *c.* 20 m a.s.l. on exposed sites which sloped slightly towards the south on a substrate that was strongly eroded, with small blocks, sandy and gravelly disintegrated basalt, and the snow cover was already gone by 12 June.

L. confusa survived the winter with a pair of partly green leaves emerging from the shoots. On each of the leaves the green colour was preserved up to 3–7 mm above the base. On 12 June it was possible to observe fresh leaves on each of the shoots. They were green with a brownish tip and 2–5 mm long. These leaves were protected by old, dead leaves, the length of which amounted to 2–5 cm. In addition, it was possible to see that the regenerative shoot was elongating. It had survived the winter alive and carried immature buds. These shoots reached up to the top of the dead leaves or slightly above them. On 15 June many new vegetative shoots were observed emerging from the leaf sheaths. They were 1–3 mm long, the youngest whitish, the more advanced ones brownish.

On 1 July the length of the youngest green leaves on the overwintering shoots had reached 10–11 mm and on 6 July these leaves were up to 19 mm in length. On 6 July the inner, leafy rosette had sprouted regenerative shoots with black buds. Here and there the peduncles were slightly elongated and inflorescences reached up above the rosette. On 9 July the first signs of anthesis appeared; anthers could be seen. Some of the inflorescences were sessile, others had grown to 12 mm tall. The new vegetative shoots had reached 10–25 mm in length. On 23 July flowering was over; the height of the pedunculate shoots did not exceed 20 mm (the length of the peduncles was 6–7 mm). On 6 August capsules formed. By 19 August the plants were ready for the winter, i.e., the leaves and the regenerative shoots had taken on a brownish colour. The ripening and dispersal of the seeds apparently occurred only during the following year (no definite observations were made).

Caryophyllaceae

Cerastium arcticum Lge. is a cushion-forming amphiatlantic species with a mainly arctic distribution area. Outside the Arctic it occurs only in the mountains of Scotland and Scandinavia (see p. 53). On Zemlya Aleksandra it was frequently met with in the form of small cushions, measuring *c.* 3 cm, rarely 6–8 cm, in diameter on sites where there is a thin snow cover and an average snow regime. These sites were usually free from snow not later than 22 June. *C. arcticum* was rarely found where the snow was less abundant, i.e., where the snow cover persisted only through the first half of June. It was absent from habitats with excessive precipitation and from those where the snow persisted up to 15 July. The plants were often encountered on all patches that were thawed out early (by the end of May). The bases of the youngest leaves survived the winter in a green state within the cushions.

On 2 June along the shore of Zaliv Dezhneva, on thawed-out sites at an elevation of 20–30 m a.s.l. with small blocks and a sandy, gravelly substrate, numerous green cushions of *Cerastium arcticum* could already be seen, although to a large extent the green colour was still hidden under the yellowed leaves from the previous year. On the lower portion of the cushion, pressed flat to the ground, about three pairs of green leaves were observed growing out from shoots. They had reddish tips. On the top of the shoots two pairs of fresh leaves could be seen. On 15 June the length of the fresh leaves had increased and they were already slightly longer than the old, dead leaves. In the lower part of the cushion, covered by fresh leaves of the current season's growth, sessile, reddish, hairy and compact buds were found on 1 July. On 6 July the buds on the majority of the cushions were still closed although swelling.

There are two stages during the bud stage: compactly closed buds, where the corolla is completely covered by bud scales, and loosening buds, where the corolla can be seen but the bud has not yet broken fully open (Shul'ts, 1966).

In one of the cushions on a slope that was exposed towards the south the beginning of anthesis was observed on 6 July. It was taking place in the lower portion of the cushion, which was pressed to the ground and facing south; a flower had started to open and the petals projected by 1 mm between the sepals. This flower was sessile and did not reach above the leaves of the cushion. On 9 July three flowers were almost open, and there were four half-open buds in the same cushion; the tips of the petals on the half-open flowers reached 1–2 mm above the leaves of the cushion. On the same day there were signs everywhere within this stand (relevé no. 20, see p. 93) of the start of a profuse anthesis. In other habitats, where the snow cover persisted longer, the cushions of *C. arcticum* were then either at the vegetative stage or at the bud stage.

New flowers continued to appear up to 10 August, and those which had started earliest and were already at the post-anthesis stage had grown peduncles up to 4 mm long, on which the beginning of capsules could be seen. On 20 August the capsules were still not ripe, were still coloured green and the majority of them protruded only 5–6 mm above the leaves of the cushion so that the base of the calyx remained covered by leaves. Simultaneously it was observed that yellow capsules from the previous year were opening and round, black seeds could be seen within them. Thus, the ripening and the dissemination of the seeds of *C. arcticum* also take place during the year following that in which the seeds were set.

Cerastium regelii Ostenf. ssp. *caespitosum* (Malmgr.) Tolm. is a high–arctic taxon of a circumpolar species. Its distribution area is basically amphiatlantic but it reaches also into the Siberian Arctic. On Zemlya Aleksandra it is met with in the form of compact, always sterile cushions, usually measuring 3–4 cm, rarely up to 8 cm in diameter. It avoids the driest localities with little snow and also those habitats where snow cover persists up to 15 July.

According to information from Tolmachev (1971), the sterile cushions of *C. regelii* ssp. *caespitosum* have the ability, in different parts of the Arctic, to split up and be carried away by melt-water, which can serve as one manner of dispersal in this species. Apparently, the same also occurs on Zemlya Aleksandra. No direct observations of this were made.

Minuartia rubella (Wahlenb.) Hiern is a cushion-forming, circumpolar arctic–alpine species. It was found once in the form of a compact cushion, 3 cm in diameter, near Zaliv Dezhneva at *c.* 30 m a.s.l., where the snow had disappeared by the end of May. It was growing on a substrate consisting of gravel, sand and strongly eroded disintegrated basalt.

The observation of this cushion of *M. rubella* was started on 2 June, when greening had already started due to the sprouting of fresh green leaves with reddish hairs along the margins. On 6 July compact buds were observed and on 23 July the sepals on one of the buds had opened by about 5 mm. On 7 August there were two flowers on the cushion; in one flower the petals projected 2 mm, in the other, 1.5 mm. The flowers were sessile. By 20 August the flower which opened first had wilted; two more flowers were still in bloom and were half-open and sessile. No capsules from the previous year were observed.

Stellaria edwardsii R. Br. is a plant with long roots and shoots which trail over the ground. It is rhizomatous and semi-creeping. The area of distribution is circumpolar–arctic. On Zemlya Aleksandra individual specimens were met with growing on sites where the snow had gone before 15 July. This species grows in the form of loose patches, rarely reaching 20 cm in diameter, forming sheets pressed to the ground and enclosed within dead, yellow leaves from the previous year. Green leaves appear only here and there at the tips of these shoots.

The plants emerged from below snow patches which thawed out early, i.e., at the end of May. In such patches, thawed out along Zaliv Dezhneva, green, slightly reddish leaves were already observed on 2 June at the tips of shoots surviving among dead, yellow leaves; they were still covered by leaves from the previous year. By 15 June the new, glaucescent leaves protruded 4 mm from below the old ones. On 19 June the green leaves were already clearly visible above the leaves from the previous year. Closed flower buds were discovered on various shoots.

On 22 June abundant new shoots, 0.5 to 3–4 cm long and whitish at the base, were observed in the same habitat; their leaves were already as long as those which had still not reached a mature condition. Very many of the fresh shoots were growing out from the above-ground parts of the roots; a portion of the old shoots were dead. On 6 July there were up to three pairs of green leaves on the recently emerged shoots and at the tip of some of the surviving shoots compact, sessile buds covered by fresh leaves were observed. On 25 July individual ripe buds had developed and on 6 August it was possible to confirm that abundant buds were opening on plants along Zaliv Dezhneva as well as on the sandy-gravelly Glavnyy Val.

The beginning of autumn coloration was noticed on 7 August: the tips of the leaves became discoloured and the tiny mats were losing their green hue. By 25 August the buds, including those not yet open, had dried up and the turf had turned brownish-yellow. Capsules from the previous year were not observed on any of the specimens of *S. edwardsii*. It is possible that its dispersal on the island occurs by vegetative means, e.g., as a result of the partitioning of the root system or by the dispersal of small, axillary buds, which could be spread by winds or water, as reported from Taymyr by T. G. Polozova (personal communication).

Papaveraceae

Papaver polare (Tolm.) Perf. is an almost circumpolar–arctic, mainly high-arctic species. It grows in the form of a dense cushion-like turf, although Polozova (1978) mentions a sympodial rosette-like growth form rather than a cushion-like one. The strongly branching crown of the root forms at the end of each branch monocarpic rosette-like shoots the accumulation of which results in the above-ground portions of the plants forming cushion-like clumps. On Zemlya Aleksandra the dimensions of the sward amounts, when growing under the most favourable conditions, mostly to 3–6 cm in diameter, rarely up to 10 cm. Under less favourable conditions, especially if the poppies grow among mosses, competition with which they cannot tolerate, the result is usually depauperate specimens of lesser dimensions.

On Zemlya Aleksandra the poppies are met with in the greatest abundance on relatively dry habitats with little snow and a rocky-sandy-gravelly substrate or on a sandy-pebbly substrate where there are few mosses (cover: only 1–12% of the surface), represented by small species such as *Bryum rutilans* and *Encalypta alpina*. The former types of habitat occur in the area adjacent to Zaliv Dezhneva, the latter on the top and upper portion of the slopes of Glavnyy Val. Under such conditions the poppies are not only abundant (Tables 8, 12) but able to produce ripe seeds as well. In other localities, where the snow cover disappears at a reasonably early date, some specimens flower but the seeds do not ripen, and many cushions remain in a sterile condition. Poppies are not found in localities where the snow persists until after 15 July or in habitats which are very wet.

The poppy cushions first started to turn green (i.e., reached a stage where the green colour of the fresh leaves began to cover the brown colour of the previous year's leaves) in localities that became free from snow at an early date, e.g., 15 June. The first flower buds – sessile, closed and greenish-red with black hair – were observed on 1 July. A mass appearance of these took place during the first ten days of July in habitats that were most favourable for poppy growth. On 19 July the first flowers were observed. A mass flowering of *P. polare* occurred between 22 July and 10 August.

The successful development of the buds was favoured by the striking capacity of the poppy peduncles to react thermotrophically such as I have described previously in respect of plants of this species in the arctic tundra (Aleksandrova, 1960). On Zemlya Aleksandra the peduncles of the poppies, reaching above the leaves in the cushion, arch at first in a hook-like manner so that the buds do not emerge above the surface of the topmost leaves in the cushion, the air within which is warmer thanks to direct heating by solar radiation. As they grow, the peduncles continue to arch, so that the buds, when the stalks lengthen, will gradually slide up to just below the surface of the cushion but not above it. As a result of this the peduncles become loop-shaped or snake-like. If the buds are lowered from the surface of the cushion to the ground, they will touch stones, pebbles or large rocks. Just after the flowers have opened the peduncles raise them slightly, after which the period during which the flowers remain in a raised position varies depending on the fluctuation of the air temperature. Even during the warmest days the flowers are not elevated more than 3 cm above the ground.

At the beginning of August the capsules of those flowers which had opened first began to ripen. Because the development of the peduncular shoots on individual poppy cushions occurred at different times (just as in the tundra zone), some specimens were observed which had ripening capsules, flowers, buds that were still opening and buds that were at the closed stage just before opening. Some flowers were seen on the last day of my observations, i.e., on 3 September, when the ground had frozen and was covered with rime. Some were also buried beneath the snow. Consequently it appears that cushions emerging from below the snow in the spring carry capsules in addition to the new shoots (in the most favourable conditions of growth) as well as shoots with ovaries, which had not succeeded in forming capsules the previous year. I also observed shoots carrying flowers which had disappeared under the snow during the preceding autumn. These flowers emerged from below the snow both withered and shrivelled at the same time as well-developed, vigorous shoots carrying capsules appeared; these latter continued to do well into July, when the capsules had still neither wilted nor stopped producing seeds. Thus, seeds of poppies are also able to ripen during the second year. Capsules were observed to start opening on 6 July but the mass production of seeds occurred only during the last twenty days of July.

It was interesting to note the difference in behaviour of poppies in two different areas on Zemlya Aleksandra where the conditions for their existence were most favourable, i.e., in habitats with a scanty snow cover near Zaliv Dezhneva and on the top and slopes of the Glavnyy Val. In the area of Zaliv Dezhneva during my year of observations there were many specimens which had produced abundant capsules the previous year. Some began to ripen on 6 July during the study year but a mass opening of these capsules and the dissemination of seeds occurred around 20 July. At that time the majority of the poppy cushions in the area of Zaliv Dezhneva had not set buds, and although some of them eventually flowered, the poppy anthesis was comparatively limited there. Apparently the populations at Zaliv Dezhneva had not accumulated adequate nourishment for a new regenerative cycle following the abundant seed set of the previous year. At the same time, on the sandy-pebbly substrate of Glavnyy Val, the poppies flowered abundantly during the year of observation but had few capsules with seeds from the previous year. Between 16 and 26 July a relatively warm period of weather occurred (Fig. 51), and on the top and upper portions of the slopes of the Glavnyy Val, the development of the generative organs of the poppies was exceptionally active. On 22 July 48 buds were observed on a single specimen, and during the period of mass flowering there were more than 20 flowers on several of the cushions.

In the case of the poppies autumn coloration, i.e., reddening of the tips of the leaves, started to develop on 5 August. Later the leaves became still more intensely coloured and finally took on a hue of a brownish shade.

Let me summarize my observations on some different cushions of poppies:

1. A specimen of *P. polare* on a patch in the area of Zaliv Dezhneva which thawed out early had, on 2 June, four green leaves, 8–10 mm long, which were covered by many long, dead leaves from the previous year; the green leaves were on shoots forming rosettes. On this cushion I counted also five fertile shoots from the previous year which were carrying capsules that were not yet ripe and covered with black hairs. On 12 June there were five green leaves, 14 mm long, on the shoots. On 15 June, after three days with positive temperatures, the cushion had turned green: the green colour hid the brown colour of the previous year's leaves. There were 5–6 new, fresh leaves, 14–15 mm long, on the shoots. On 1 July the cushion was still in the same condition. On 6 July the length of the fresh leaves had reached 16 mm, on 23 July, 18 mm. Of the capsules from the previous year four had ripened and were dispersing seeds, the fifth had cracked open and dark brown seeds could be seen inside it. The observations of this cushion extended to 20 August, but in that year (1959) flower buds had not yet appeared on it by then.

2. A poppy cushion growing among pebbles on the northern slope of Glavnyy Val had, on 5 July, two closed, greenish-red sessile flower buds with black hairs, one 4 mm in diameter and the other 6 mm. On 7 July there were already four sessile buds from 4 to 9 mm in diameter. The hairs were thicker on the buds which appeared first and they had acquired a black colour. On 9 July there were six buds on the cushion; the ones first appearing had elongated peduncles (maximum length of bud and stalk: 16 mm). On 11 July the most developed peduncles were 27 mm long. They were arched so that the buds were resting inside the top portion of the cushion and not showing above the tips of the leaves. On 13 July thirteen buds were noticed on this cushion. By 17 July the length of the longest peduncles had reached 30 mm, of which 9 mm were taken up by the bud itself. The stalk was curved so that the bud did not appear above the tips

of the topmost leaves of the cushion. Five of the 13 flower buds had narrow, reddish cracks between the sepals; on one 1 mm of the yellow petals could be seen. Because of the elongation of the peduncle, one of the buds had been pushed down from the cushion and was resting adjacent to a row of pebbles, absorbing additional heat from the stones during the daylight hours; see p. 158 concerning a similar case of *Poa abbreviata*. On 20 July the length of the best developed regenerating shoot measured 34 mm (including 9 mm for the bud itself). The yellow petals could be seen peeping out of the buds through cracks, 1–1.5 mm wide, between the sepals. On 1 August the length of the peduncular shoots exceeded 50 mm; all were arched so that none of the buds appeared above the tips of the leaves or so that the bud rested on nearby stones. One half-open flower was observed – its petals had spread 4 mm apart and the corolla reached 4 mm above the tips of the leaves in the cushion – and four buds were loosening up but the rest were still closed. On one of the buds the petals were closed but could be seen protruding 5 mm beyond the sepals. All the peduncles, flowers and buds were pressed to the surface of the cushion or the adjacent stones. On 10 August the flowers developing first had wilted. On 15 August the petals began to fall off the first flower (two petals were still hanging on). Another two had finished flowering, four were seen in full bloom and there was one flower with a half-open corolla. Two flowers were somewhat (0.5–1 cm) elevated above the cushion (it was an exceptionally warm day (see Fig. 51)) but the rest touched the tips of the leaves in the cushion or the adjacent stones, as before. On 24 August the four flowers with their capsules swelling behind a thick cover of black hairs (at this stage, the capsules were almost spherical) had dropped their petals; one of the capsules was lying against a pebble, the rest were raised 1.5–3 cm above the ground. One flower had reached anthesis, three were still in full bloom. On 28 August it rained and there were strong north-easterly winds. The petals of all the flowers on this cushion fell off and were stuck to the stones by the rain, and all the capsules except one were broken off. However, the majority of the flowers and capsules on poppy cushions that were better protected from the winds survived, especially those growing on the south-facing slope of the bank.

Cruciferae

Cardamine bellidifolia L. is a cushion plant with a circumpolar–arctic area of distribution. It is rarely met with on Zemlya Aleksandra and then always growing as separate individuals, exclusively in habitats with optimal dates for the disappearance of the snow (between 10 and 22 June) and with the most favourable soil and substrate conditions (Tables 7 and 8). Single specimens can often be found on fine soil with rocks or on pebbly substrates where the snow cover has gone by 1 July. None are ever seen on sites where the snow disappears later. Usually the plants grow in the form of small cushions with a diameter of 1.5–2 cm, rarely 5 cm.

There were only closed buds on all the specimens of *C. bellidifolia* observed at the beginning of July. The buds had started to loosen up by 5 July. The first flowers were noted on 9 July; a mass flowering was observed on 25 July (all the flowers were sessile), and by 5 August the flowers had reached the post-anthesis stage and siliquas were beginning to form. The best developed siliquas were slender, short (*c.* 5 mm long) and of a blackish-violet colour. From lack of observations it was impossible to find out whether the seeds of *C. bellidifolia* ripen during the year in which they are set or the following year. Siliquas from the previous year, 5 mm long and slightly elevated above the leaves of the cushion, were noticed here and there on 20 July, but the valves had not yet opened. Autumn coloration appeared in August and resulted in the leaves turning bright orange or brownish.

My observations of one cushion of *C. bellidifolia* growing under the most favourable conditions (on top of the south-facing slope of Glavnyy Val) were as follows:

On 19 June some sessile, dark violet, closed and very small buds were noticed. On 22 June the edges of the sepals had turned just a little lighter in colour on some of the buds. On 23 June the majority of the buds were slightly whitish along the seams. On 25 June very narrow, dark cracks appeared between the sepals but the buds remained sessile as before. Between 27 June and 5 July no visible changes took place; this coincided with an exceptionally cold period of weather (Fig. 51). On 7 July the sepals on three flower buds had separated by up to 0.5 mm and white petals began to show. Between the petals a slit, tiny as a pinprick, had appeared and anthers could be seen through it. On 11 July one flower was observed opening up and many closed buds were seen (at another locality, the first open flower appeared on 9 July; on 13 July there were 30 flowers and 9 buds were opening. Below the flowers which opened up first, the peduncles had grown to 2 mm in length and the petals were therefore raised slightly above the leaves, but the tips of the sepals still remained at the same level as the leaf tips of the cushion. By 20 July 42 flowers had opened, 8 buds were loosening up and the entire cushion appeared white with flowers. On 23 July almost all the flowers had opened except for a few that were still loosening up. On 7 August the mass flowering was over; some of the petals were still white but crumpling up while others were turning grey and falling off. Slender, very small (*c*. 3 mm long) siliquas began to form from the ovaries; they were dark violet and had peduncles no longer than 2 mm. On 18 August petals were still left on some of the flowers although they were withering, turning greyish white and wrinkling. The petals had fallen off the other flowers, and siliquas, 5 mm long, could be seen, dark violet in colour, slender and looking stunted.

Draba subcapitata Simm. is a cushion-forming, almost circumpolar, high-arctic species (Fig. 52). It is less often met with on Zemlya Aleksandra than the other species of *Draba*, and flowers later. It occurs as separate individuals mainly at sites with a scanty snow cover but can also be found where the snow does not disappear until after 1 July and is all gone by 15 July. It grows in the form of very dense small cushions, the dimensions of which usually do not exceed 1.5–2 cm, rarely 3 cm, in diameter.

When just emerging from below the snow, the cushions were almost white due to the predominance of the dead, external leaves, which had a whitish colour. The rosette of leaves of which the cushion was formed were closed at the top. On 12 June the rosette of leaves started to spread apart and inside them two leaves, which had survived in a green state, could be seen together with some small, dark violet, sessile and closed flower buds, which had also overwintered. On 15 June the buds had become bluish and had swelled slightly (up to 1.5 mm). On 28 June the green leaves of the rosette were seen to be elongating, although they were still hidden by the whitish, dead leaves from the previous year.

On 1 July the first sessile flowers with white petals were observed. The cushions had now turned green: the green colour of the leaves, now increasing in length, dominated over the colour of the dead leaves. Mass flowering occurred during 5–7 July. On 10 July the plants were still flowering but were observed to be approaching the post-anthesis stage and on 15 July the petals began to fall off. (On that date other shoots were still flowering.) On 20 July some ovaries had started to swell and exceeded the length of the long perianth; on sites where the snow had disappeared at a late date, plants were still flowering.

On 5 August siliquas began to form but these were still almost completely buried within the cushion, their tips reaching above it by only 1.5–2 mm. Simultaneously, next season's very small, closed, sessile, violet flower buds could be seen on the shoots. The siliquas reached full shape between 13 and 18 August: they were almost spherical, green and glabrous, 2.3–3 mm long, and at this time exceeded the length of their peduncles by *c*. 1 mm. The siliquas were raised above the leaves of the cushion by almost their full length. They had not yet opened by the end of August. Apparently the seeds are dispersed at the start of the following season since from 23 June onwards dry siliquas with brown seeds from the previous year were observed more than once.

Draba oblongata R. Br. is a cushion-forming plant with a high-arctic, almost circumpolar area of distribution. It is one of the most (or rather) widely distributed species on Zemlya Aleksandra and is met with in all kinds of habitats where plant aggregations occur, although only as

Fig. 52. Area of distribution of *Draba subcapitata* Simmons (according to Hultén, 1968; *Arctic flora of the USSR*, vol. 7, 1975).

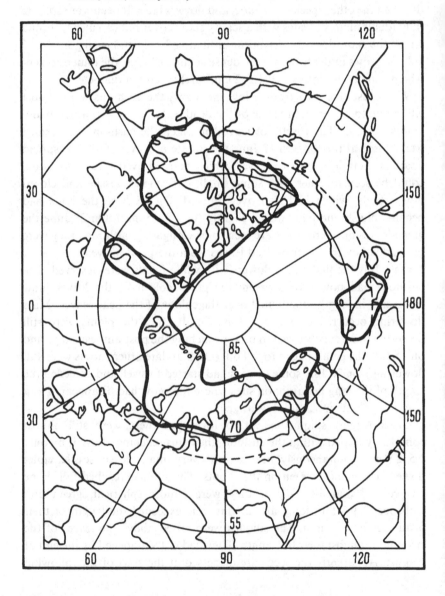

individual specimens. It is especially significant that *D. oblongata* grows as individual specimens scattered here and there in those localities where the growing period is limited by the late disappearance of the snow cover: as a rule specimens of *D. oblongata* could be found where the snow did not disappear until 20 August (Table 18) but they were also observed in habitats with a lighter snow cover. Individual cushions were often very small, *c.* 1 cm in diameter, although the dimensions of the largest ones could reach 2.5–3 cm across.

When they became free from snow, the external rosette of leaves forming the cushions were dead and greyish-brown but the youngest leaves, i.e., those inside the rosettes, were greenish. The leaves on the shoots which formed the rosettes were inclined towards the centre, their tips meeting. On the part of the shoots which had not reached a regenerative stage during the preceding year, the tips of the centripetal leaves covered closed, sessile flower buds, which were thus protected from below. On specimens of *D. oblongata* that had appeared from below snow on 1 and 2 June all the rosette leaves were still closed on 12 June and formed something similar to little 'chambers'. On 13 June the tips of the rosette of leaves on some cushions had spread apart by *c.* 1 mm. On 16 June the plants began to turn green, especially on those parts of the cushions which touched the ground. On 22 June half-open buds were noticed and masses of such buds were seen on 24 June when the first flowers were observed.

On Zemlya Aleksandra the tiny flowers of *D. oblongata* are not only sessile but also hidden deep within the rosette and flower under the shelter of the leaves forming a canopy above them. It is only during post-anthesis that their peduncles elongate a little but even then the flowers remain covered below the tips of the leaves. It was only when the ovaries started to swell, during the first 10 days of July, that the tips of the flowers began to emerge, at first by only 1 mm above the rosettes. Finally, on 20 July the siliquas were observed to be forming and their tips became raised about 2 mm above the rosettes.

On 23 July mass fruiting of *D. oblongata* was observed. Well formed, brownish-violet siliquas were almost fully exserted above the cushions or had only their bases covered by the leaves. By 25 August the majority of the siliquas had opened up. Siliquas which had opened in the autumn of the previous year could be seen on cushions of *D. oblongata* as soon as they became free from snow in the spring. Only a part of the siliquas from the preceding year had not yet opened. My observations revealed that they ripened and opened up at an early date.

Thus, *D. oblongata* displayed a capacity for passing through the regeneration cycle of its development exceptionally quickly. The duration of the vegetative period was only fifty days; its very small cushions with green but fully formed siliquas could be seen during the second half of August at sites where the snow had disappeared only between 15 and 20 July.

My observations concerning the seasonal development of one specimen of *D. oblongata*, which had become free from snow on 2 June, are as follows:

On 2 June the tips of the leaves on all the shoots of the rosette forming the cushion were closed. A portion of the shoots had compact flower buds, formed during the previous autumn. On 12 June no visible changes had taken place. On 13 June the tips of the rosette of leaves had spread apart by about 1 mm. On 16 June the cushion began to turn a greenish colour, especially in the lowest parts of rosette shoots close to the ground. On 17 June the buds were closed and sessile as before but the leaves of their 'rosette house' had spread further apart although they were still inclined towards the centre and arched over the buds. On 22 June the buds began to open up. On 23 June there were two half-open flowers inside the cushion, one of which showed the light-yellow, narrow petals; the other had 1-mm wide slits between the sepals. On 24 June the petals on the first flower had straightened out, on the other the sepals had spread apart by 1.5 mm and the petals by 0.5 mm. On 25 June no changes were noted. On 26 June the petals of the second flower were 2 mm apart. On 28 June the petals of the first flower had faded; the second flower remained in the same condition as before. By 3 July the petals had faded also on the second flower. The flowers remained sessile, hidden deep within the rosette in the shelter of the leaves inclined above them. By 7 July the peduncle of the first flower had grown a little and the flower was just 1.5 mm below the tips of the leaves. By 9 July the shoot had stopped flowering; the length of the peduncles had increased a little more and the tips of the flowers were level with the surface of the top leaves in the rosette. On 11 July both flowers had wilted and their peduncles had elongated still further, the tips of the ovaries jutting 0.5 mm above the rosette formed by the centripetal leaves. On 13 July the tips of the leaves of the cushion had spread apart by about 1 mm; the ovary was protruding slightly above the sepals, overtopping them by 0.5 mm. By 20 July 4-mm long and 2-mm wide siliquas had formed, their tips reaching 2 mm above the cushion. They sat on peduncles which were 1.5 mm long. On 23 July the siliquas were brownish-violet and fully extended above the cushion. They opened up on 20 August. The majority of the siliquas on other specimens of *D. oblongata* growing in the same habitat had also opened up.

Draba pauciflora R. Br. is a cushion-forming, arctic–alpine species. From a systematic point of view it is close to *D. oblongata*, belonging like that species to the section *Oblongatae* and having a similar biology and ecology. Although only found as individual specimens, it occurs in habitats where the snow disappears earlier than in the case of other species of *Draba*; it flowers early and passes quickly through all the stages of the regeneration cycle. Therefore, seed-bearing specimens could be seen, as could, of course, specimens with siliquas that were not yet ripe, in August in places where the snow had melted only on 15 July. However, *D. pauciflora* was more rarely met with than *D. oblongata*, and in habitats where the snow disappeared at very late dates it was rarely discovered. It grows in the form of small cushions, usually only 1–2 cm in diameter. In the most favourable habitats it can reach 3–3.5 cm and, very rarely, 4 cm in diameter. Very small cushions, only 1.5 cm wide, with siliquas formed the preceding year were frequently encountered. The cushions were found on patches that were already thawed out at the end of May and had survived the winter within the dead, external (older) leaves on shoots inside the rosettes, where they displayed four green, young leaves.

When they became free of snow, the leaves on the shoots forming the rosettes were united at their tips so that they covered the buds that were formed during the previous autumn and which were situated deep within the rosettes, just as in the case of *D. oblongata*. After about ten days the rosettes of leaves began to spread out, although they remained inclined towards the centre and continued to cover the developing flower buds. By 17 June the buds were seen to be opening up and at the end of June the first flower was observed. Siliquas began to form on 17 July and the majority of them had opened by 25 August.

My observations of different specimens were as follows:

1. On some patches, thawed out close to the Nagurskaya polar station, the youngest (inner) leaves had turned green on 2 June; the tips of the leaves were completely enclosed within the rosettes. On 13 June the leaves in the lower part of the cushion, which were adjacent to the ground, had spread 2–4 mm apart. At the bottom, inside the rosette, still completely closed, overwintering flower buds could be seen. On 16 June there were no visible changes. On 17 June three buds were noticed loosening up inside the rosette; they were covered by the leaves, bending in over them, and their yellow corollas were visible.

2. In a specimen on the upper portion of a south-facing slope of Glavnyy Val, the leaves of the shoots forming the rosette pulled apart on 12 June, although they still remained closed over the inside parts. In their shelter at the bottom of the rosette, sessile, closed, violet buds (five within this rosette) were observed. On 15 June the buds had turned bluish and

were slightly increased in size, up to 1.5 mm. On 1 July there were two sessile flowers; in the rosette, another six buds were noticed opening up, but the rest of them remained closed. On 6 July six flowers were observed inside the cushion: the blades of their petals had straightened out and the length of the peduncles had increased slightly so that the tips of the petals were almost at the same level as the tips of the leaves. Small flies were stirring about on the flowers. By 17 July siliquas, 3.5 mm long and 2 mm wide, had formed, their tips reaching 1 mm above the leaves. By 28 July the siliquas had become 4–5 mm long and protruded up above the cushion by their full length. Those siliquas which had formed earlier than the rest opened up at the end of August but those which had formed later remained closed and were covered by snow, still unopened, so that they could ripen and open the following year.

Draba macrocarpa Adams is a cushion-forming species with an almost circumpolar–arctic area of distribution (Fig. 53). It is not often found on Zemlya Aleksandra, where it grows as individual specimens and is considerably rarer there than the other species of *Draba*. It thus mostly occurs in habitats where the snow disappears at the optimal time; occasionally it is also encountered at sites where the snow does not go until after 1 July but where it is gone by 15 July. The cushions and the siliquas are larger than those of the other *Draba* species found on the island: the predominant size of the cushions is 4 cm (1–7 cm) in diameter.

As in the case of the species described above, the multi-leaved rosettes of this *Draba* were, when emerging from below the snow, closed up, and at their centre there were closed, sessile flower buds that had been formed during the previous autumn. Later on, the leaves started to spread apart but continued to bend over the interior, sheltering the developing buds. By 20 June the appearance of buds loosening up was noticed and the first flowers were observed on 24 June. On 10 July the siliquas began to form. A portion of the siliquas ripened and opened at the end of August, another portion remained closed over the winter and ripened the following year. At sites most favourable for occupation by *Draba macrocarpa*, especially on the slopes of Glavnyy Val, where it develops best in 'stone greenhouses', cushions with a large number of siliquas could be seen. The siliquas observed in the spring had survived the winter; those observed in the autumn had been formed the same summer. All the siliquas were either elevated by their full length (6–7 cm) above the cushions or just had their bases covered by the tips of the leaves. At the start of the season only few of the siliquas were open, i.e., only those which had ripened the preceding autumn; during the first ten days of July those seeds which had set the year before concluded their ripening and cushions with a large number of the previous year's siliquas opening and the light brown seeds being dispersed from them were encountered.

My observations concerning the seasonal development of different specimens of *Draba macrocarpa* on the upper portion of the south-facing slope of Glavnyy Val (under the most favourable conditions) were as follows:

On 22 June the flower buds – four on each shoot – were still closed, sessile and covered by the rosette, the leaves of which had already begun to spread apart although they still remained inclined towards the centre. On 23 June one bud appeared to be loosening up and on 26 June two more

Fig. 53. Area of distribution of *Draba macrocarpa* Adams (according to Hultén, 1968; *Arctic flora of the USSR*, vol. 7, 1975).

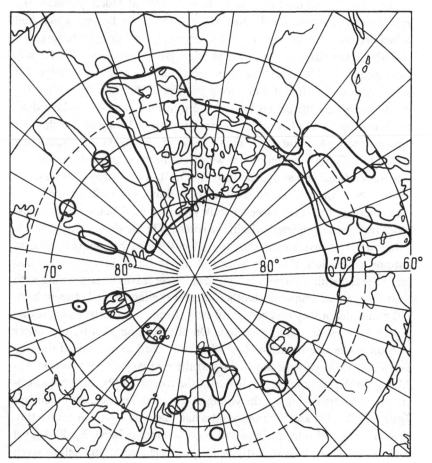

appeared to be doing so. On 28 June the petals could be seen spreading about 1 mm apart on the first bud and *c.* 0.5 mm on another; four of the other buds were seen to be loosening up while the rest remained closed. All the buds were still sessile. On 29 June two more buds appeared to loosen up. On 30 June three flowers were observed to be opening up and six buds to be loosening. By 4 July the petals of the first flower had paled (the start of the post-anthesis stage). By 5 July the petals had faded also on the second flower. All the flowers and buds were still sessile: the tip of the flower that had bloomed first was about 3 mm below the tips of the leaves. By 9 July many flowers in the cushion had wilted but remained sessile. On 11 July the majority of the flowers were finished and the peduncles had started to elongate: some of the ovaries already reached the upper edges of the rosette leaves. The last of the siliquas retained on the cushion from the previous season opened and the light brown seeds could be seen. On 13 July the tips of some flowers, already faded, projected 0.5 mm above the sepals. On 17 July the ovaries of the flowers which had wilted first were 5 mm long and 2–2.5 mm wide, their tips protruding by 2 mm above the cushion. The siliquas were green or reddish. On 20 July large siliquas had formed; these were violet and their tips were raised 4–5 mm above the cushion; a 2-mm long portion of the stem supporting the inflorescence and the peduncles, 1.0–1.5 mm long, was still hidden deep inside the rosette. On 1 August eight shoots carried 2–3 siliquas each, their tips protruding 4–5 mm above the cushion. On 6 August the siliquas had grown to 6–7 mm long and 3 mm wide, had become greenish-brown and were raised 6–6.5 mm above the cushion. On 10 August they were elevated further above the rosette and some of the peduncles (now 2 mm long) had also emerged; the small piece of stalk (about 5 mm) below the inflorescence remained completely within the rosette. On the shoots that were not going to flower that season, the external leaves were beginning to take on autumn coloration; the four youngest inside leaves in the rosette had turned green with a reddish-brown hue. These leaves were growing with their tips pressed against each other; there was either a gap between them, 1–2 mm wide, or they touched completely. Inside this rosette small, sessile flower buds were again observed. On the fruiting shoots, the upper portion of the external leaves had wilted one third to one half of the way down. On 15 August the dimensions of the siliquas had not changed. On 24 August the lower (external) leaves of the rosette had become brownish-yellow and were drying up; the youngest (internal) leaves leaned their tips against each other and their outer sides had taken on a greyish-brown tint. Two siliquas had split, the rest had become more inflated. By 30 August some siliquas had turned partly yellow and their septa and yellow seeds could be observed. The majority of the siliquas remained unopened; they would ripen and split open only during the following season.

Cochlearia groenlandica L. is a monocarpic rosette plant with a circumpolar–arctic area of distribution. On Zemlya Aleksandra the largest number of plants of this species was found at Zaliv Dezhneva (Table 12) growing at 20–30 m a.s.l. in habitats with small blocks of a substrate of gravelly disintegrated basalt, where the snow disappeared early, generally at the end of May. It was less abundant in other habitats where there is a scanty snow cover, but it is not absent from sites with deeper snow either: it grows there in the form of individual specimens only. It was once discovered, also as single individuals, in a mossy 'bog-like' habitat (Table 21), where some of its rosettes succeeded in flowering by 7 August.

In the most favourable habitats *C. groenlandica* exists in the form of populations containing individuals of different ages, from seedlings and young plants to a stage in the life cycle where they have entered the regenerative phase. They grow in the form of very small rosettes, flattened to the ground. The diameter of rosettes at the vegetative stage does not exceed 2.5 cm, but when flowering it reaches 4 cm and at the fruiting stage it can, under certain conditions, increase to 7 cm thanks to the growth of the regenerative stems, which are also pressed flat to the ground. The length of the leaves is about 1 cm, including *c.* 4–5 mm for the leaf blade itself.

The observations started on 2 June in the area around Zaliv Dezhneva. On that date one of the *C. groenlandica* rosettes (a number of which were labelled) had two external rows of white, dead leaves, then three rows of overwintering leaves, all yellowish-green with the blades overlapping like roof tiles. The central portion of the rosette was occupied by a large quantity of very small buds, formed during the previous autumn and completely unprotected. By 12 June the buds had enlarged slightly, to 1 mm in length, and the leaves had become green. By 15 June the buds at the peripheral row of the inflorescence had spread 4 mm apart and a portion of them had turned whitish, but petals could not yet be seen. The leaves were finally green all over. On 19 July a portion of the flowers at the periphery of the inflorescence were half-open. In the same habitat there were already fully opened flowers around the rim of the inflorescences on other specimens of *C. groenlandica*.

On 1 July all the flowers around the edges of the inflorescence were in bloom. The inflorescences were assembled in a horizontal position, very small, and tightly bunched, with 6–8 flowers each on miniature peduncles, which had grown out by this date, and which had a single, tiny stem leaf placed near the inflorescence. The length of the peduncles was at most

12–15 mm in the outside row and decreased gradually from the edge of the rosette towards its centre. At the centre of the rosette buds on very shortened peduncles could be seen, still not open or just about to open. By 6 July all the flowers had unfolded (just as they had on the regenerating shoots of other specimens in the same habitat). They emitted a strong scent of honey and on the ground a small number of flies were flying about; a single individual had crawled up onto a flower. On 9 July this specimen was still in bloom, with the flowers still emitting the honey scent and flies crawling on them. On 14 July all the flowers were still in full bloom but a part of the petals at the edge of the inflorescence had faded.

On 23 July the petals still remained on all the flowers but the majority of these were already seen to be growing thick, chestnut-brown, almost spherical ovaries, 3 × 3 mm. Specimens of *C. groenlandica* growing at the same site but which had flowered somewhat earlier were already at the fruiting stage: their petals had faded or fallen off, their leaves were drying up and all the rosettes, completely pressed to the ground, appeared to consist of a mass of inflated, brown siliquas.

On 7 August almost all of the flowers on one of the specimens of *C. groenlandica* studied at this site were observed to have formed siliquas, although a few flowers at the centre of the rosette still had white petals. On 19 August the regenerative shoots had grown to 4–4.5 cm long, the leaves were drying up and the siliquas had started to open.

On those specimens of *C. groenlandica* which had passed the generative stage, 12–18 rows of leaf scars could be seen. Counting the annual production of leaves, it could be concluded that the age of these individuals was not less than 4–6 years. At this time they had succeeded in forming 'thick' (in relation to such tiny plants) roots, *c.* 1 mm wide, and still thicker stems, which were shaped like truncated cones. At the level of the roots the stems had a diameter of *c.* 1 mm but widened in their upper part to 3–4 mm, reaching 5–6 mm in length. The material accumulated over 4–6 years in the blades of the leaves was fully used up during the formation of the buds in the autumn and during the following year for flowers and fruits. On fruiting specimens the leaves were dry and the perennial 'vessels' of the stems and the root had shrivelled and wrinkled.

Thus, under the harsh conditions of the polar deserts monocarpic *C. groenlandica* displays a striking adaptation by the development of fully efficient populations of different ages.

Saxifragaceae
Saxifraga nivalis L. is a short-rooted, circumpolar arctic–alpine species. On Zemlya Aleksandra it was met with only rarely and growing as

isolated individuals on small blocky and on gravelly and sandy-pebbly substrates only in habitats where snow did not disappear until 15 June. It forms loose tussocks, composed of shoots with thick stems pressed to the ground and covered in the black remnants of dead leaves. The best developed and oldest stands can reach 7 cm in diameter but are generally only 3–4 cm across.

On the portion of the shoots forming the tussock, the leaves are arranged like rosettes. However, they do not spread horizontally but are directed straight up, forming a wide funnel-shaped cavity of overlapping leaves. On each shoot four interior leaves, formed the previous year, survived the winter still green, but had red tips and backs. Fresh leaves started to develop as soon as the stand became free from snow: they grew very slowly throughout the entire season and reached full size only in the autumn. As these grew and became established, the overwintering leaves gradually began to dry up, from 10 July onwards.

On some of the shoots inflorescences developed from sessile buds laid down during the previous autumn. The first flowers, still sessile and hidden deep within the funnel-shaped cavity formed by the leaves, were observed on 15 July. After the flowers had appeared, the peduncles began to elongate slowly, reaching about 10 mm in August at the same time as anthesis was completed and the capsules began to set, so that the inflorescences became just slightly (1–2 mm) raised above the rosette. No ripe capsules were discovered, maybe for lack of observation. However, on Zemlya Aleksandra the seeds of this species seem to ripen only during the most favourable years.

Flowers did not develop on the majority of the shoots. Some of these remained in a vegetative condition until late in the autumn; on others sessile buds, hidden at the bottom of the leafy funnel-shaped rosettes, were established at the end of the season.

My observations on the seasonal development of rosettes of *Saxifraga nivalis* growing on the upper portion of a slope on Glavnyy Val (a most favourable habitat) were as follows:

On 25 June rose-coloured sessile buds established the preceding autumn were observed on two shoots in a stand. The interior leaves of the rosette had survived the winter alive and were green above and red on the back. On 26 June a 0.5 mm wide split appeared between the sepals of one of the buds. By 1 July the length of the leaves had increased to 6 mm but there were no visible changes in the rest of the rosette. On 7 July all the buds were still sessile, and all were closed except one, where the sepals had pulled 0.5 mm apart. The tips of the buds were about 5 mm below the tips of the leaves of the rosette; the length of the leaves was now 8 mm. On 9

July the same leaves were 9 mm long; nothing else had changed. By 11 July yet another bud had spread its sepals apart by 0.5 mm. The overwintering leaves were turning slightly greyish-brown. By 13 July the sepals on one bud had spread 1 mm apart, on three others in the same inflorescence there were cracks between the sepals, measuring 0.5 mm; all the buds were still sessile. On 17 July one of the buds broke open: the sepals had spread 3 mm apart and between them greenish-yellow petals could be seen. These were slightly shorter than the sepals and the somewhat protruding stigma. In the rest of the buds the opening between the sepals had grown to 1 mm wide. The peduncles started to elongate. The inflorescence had become 2 mm taller than before. On 20 July four flowers in this inflorescence were open; their petals did not exceed the length of the sepals, which were slightly bent back and whitish. On 10 August the flowering stage began to wane: the petals were still white, but fading; the ovaries could be seen, knot-like, dark red and projecting *c.* 1 mm past the perianth. The height of the peduncles, including the inflorescence, was 9 mm. The leaves, which had started to develop during this season, had lengthened to 10 mm. The interior leaves of the rosette had started to turn yellow or brownish. On 18 August the height of the peduncle, including the inflorescence, was still only 10 mm, the petals had wilted and were greyish-brown and the ovaries were dark red, their tips about 1 mm above the perianth. On 30 August the external leaves of the rosette were chestnut brown and drying up; the length of the leaves, developed that season, was now 13 mm, their width 10 mm. They were green on the upper side of the blade, bright red on the lower side; the ovaries were fully mature, dark red to lilac. The height of the flowering shoot remained 10 mm, the inflorescence very slightly overtopping the rosette leaves by 1–2 mm. In this condition the stand became covered with snow.

The condition of another plant flowering in this year (1959) was the following: on 10 August closed, small, dark red buds could be seen deeply situated at the bottom of the almost cylindrical cavity and surrounded by leaves that were not yet developed; they had been prepared for the following year. The cavity was formed by four short (3–4 mm long) interior leaves with hairy edges; the adjacent six leaves were overlapping each other and formed the funnel-shaped upper portion of the rosette. The four outside leaves had brownish edges, the interior ones were green above and bright red below.

Saxifraga tenuis (Wahlenb.) H. Smith is a short-rooted, circumpolar-arctic species. On Zemlya Aleksandra it is very rare and met with only in localities that are free from snow at an early date where it forms small, loose mats, not more than 2 cm wide, consisting of rosette-forming shoots generally similar to those of *S. nivalis*. However, the leaves and the shoots are smaller and thinner than those of *S. nivalis*. On 14 August they were observed with the flowers protruding somewhat above the leafy rosette.

Saxifraga cernua L. is a circumpolar arctic–alpine species (Fig. 54) with underground stolons. On Zemlya Aleksandra it forms more or less loosely developed clumps. These can reach 4–5 cm in diameter under the most favourable conditions but they are frequently met with as depauperate individuals only 2–3 cm in diameter. When growing in habitats where the snow disappears late, they can be still smaller.

S. cernua belongs to the group of plants which are almost eurytopic on Zemlya Aleksandra (or are rather widely distributed). It is also met with as single individuals in all kinds of habitats where there are plant aggregations: it grows where the moisture regime is variable and in places with widely different dates for the disappearance of the snow, from those with minimal snow cover where the snow disappears early to those which are free from snow only by 20 July (Table 20). It survives the winter with

Fig. 54. Area of distribution of *Saxifraga cernua* L. (according to Hultén, 1968).

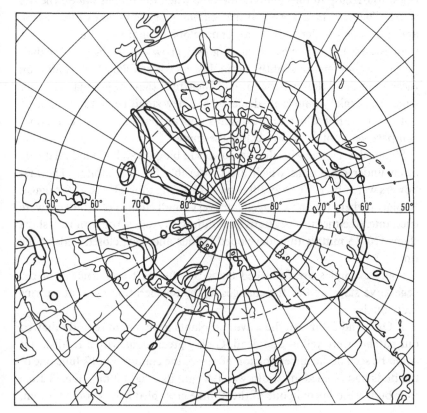

completely dead leaves on the vegetative shoots and leaf buds, developed during the previous autumn in the lower part of the clump, from which new leaves begin to sprout as soon as the plants emerge from beneath the snow.

Vegetative propagation of *S. cernua* occurs not only at high latitudes, but also much further south: when the top flowers are shed, the capsules do not set and propagation follows by means of bulbils in the axils of the stem leaves.

On Zemlya Aleksandra the stems of plants which do not carry bulbils can be 2–3 years old, as shown by repeated observations. Stems proven to be two years old (i.e., having survived one winter) can, under favourable circumstances, succeed in producing flowers during the second year of their life. Those stems on which the top buds do not succeed in opening are covered with snow again together with their leaves and rudimentary bulbils and the unopened buds that are already established on the stem. During the next season the buds that survive the winter do not open but the bulbils ripen and fall off, after which the stems dry up. New stems begin to develop in the clump at the time when the bulbils fall off the old stems. The new stems appear at the beginning or the middle of the season and are buried under snow in an undeveloped condition when not more than 10 mm tall. After overwintering they grow out to the limit of their size, i.e. 20–40 mm (or slightly more). Leaves with axillary bulbils form on these stems, and flower buds form at their tips: these either mature, after which the axillary bulbils are detached and new stems begin to form, or do not mature, after which no new stems appear that year but the old ones will pass yet another winter below the snow.

Stems of *S. cernua* with the top flowers in bloom can be found relatively often under conditions which are favourable for the habitat. The first flowers were observed on 29 July, mass flowering took place on 20 August. In places where the snow stayed late, very small clumps of *S. cernua* were encountered either in a vegetative condition (in August) or with very short stems (10–15 mm) with undeveloped bulbils and tiny, closed buds at the tip.

Some examples are furnished below of the start and renewal of the vegetative stage and of the changes in the regeneration cycle of various individuals:

1. On 1 and 2 June all the leaves on plants growing on recently thawed out patches were yellow and dead. Glossy green buds with a reddish tint could be seen in the upper portion of one of the clumps. On 12 June one of the buds had opened: its width was 1.5 mm, its height 3 mm. On 13 June four buds were open: in one of them three crumpled leaves could be seen,

reddish green, 3 mm long and 2 mm wide. On 14 June there were four crumpled leaves, 4 mm long; the longest of the leaves had begun to straighten out by 15 June and reached 2–2.5 mm across. On 17 June eight leaves unfolded for half their length were noticed. There were twelve leaves, half unfolded, 8 mm long and up to 2.5 mm wide on 18 June; they were completely covered by the dead leaves from the preceding year.

2. A clump of *S. cernua* on the north-facing slope of Glavnyy Val. On 20 June this was yellow in appearance because of the dead leaves on it. Among these an overwintering (two year old) stem could be seen, 40 mm long, with wilted leaves, bright red bulbils in their axils and slightly opening flower buds at the tip. The petals had pulled 0.5 mm apart. On 25 June the flower buds were in the same condition and so was the stem; inside the clump there were crumpled, still not straightened out green leaves, but they were completely hidden by the dead leaves from the previous year. On 29 June the buds and the stem still remained unchanged but many of the fresh leaves had unfolded and grown a little. They could be seen among the old leaves but had still not reached their full size. By 1 July half of the basal leaves had unfolded and reached 6–7 mm in length. The stem remained as before. On 5 July all the leaves were completely unfolded and had reached 9 mm in length, the clumps had turned green and the old material was now covered by the green leaves. The stem stayed as tall as before and the buds on it were in the same condition as before. On 7 July some bulbils had fallen off and were lying on the ground at a distance of about 2 cm from the stem. By 9 July the stem was both dry and broken. On 13 July a new stem had appeared at the base of the dead stem and was already 7 mm long. At the tip of the new stem there were tightly crowded leaves of a longish shape, looking like loose buds. On 17 July the new stem was 8 mm long and it was completely covered by leaves. On 20 July no further changes were found. On 1 August a new stem was perfectly hidden below the leaves, but on it the tiny beginnings of bulbils could already be distinguished in the leaf axils crowded at the tip of the stem, and at the very end of the stem tip there were small, closed flower buds. On 6 August the leaves had turned yellowish along the edges (i.e. they had taken on autumn coloration); the rest was unchanged. By 15 August the new stem had grown to 10 mm long. By 29 August the leaves of the clump had died but the stem remained unchanged.

3. On 19 June a mat of *S. cernua* on the raised shore of Zaliv Dezhneva was more advanced in its development than clumps described above because this area had thawed out earlier than other parts of the island. There were already fully developed leaves and twelve stems, which had survived one winter. Most of these stems were the same size, *c.* 15 mm

long; they had apparently already elongated after the disappearance of the snow. They had oblong leaves near their tips as well as closed, red flower buds. On 1 July the stems were *c.* 3 cm tall and at the tip of each there were closed flower buds with bulbils in the leaf axils, some of which were of a mature shape, although not those closest to the tip. On 9 July the sepals of one of the flower buds had spread 0.5 mm apart. On 14 July the stems were 38 mm in length but remained otherwise in the same condition. On 23 July one flower bud was half open; the petals overtopped the sepals by 2 mm but were still closed. At the same site there were other mats of *S. cernua* with open flowers. On 27 July more half-open buds appeared on the specimen studied and the number of specimens of *S. cernua* flowering at the site in question was growing. On 29 August the stems of the mat studied were 49 mm tall and alive with bright red bulbils, some of which were loosening and some of which remained compact. All the stem leaves as well as the basal leaves had turned yellow. The stems were buried under the snow with the bulbils still hanging on. No signs of new stems were observed in that mat.

Saxifraga hyperborea R. Br. is a circumpolar–arctic, mainly high-arctic, species. The plant produces many rosettes and forms a loose mat. Like *S. cernua*, it is one of the more widely distributed plants on Zemlya Aleksandra and is met with, although as individual specimens only, in all types of aggregations, except the 'mossy bog-like', on various kinds of soils and under all kinds of snow regime. It grows in the form of very small, loose mats, usually 1–1.5 to 2.25 cm in diameter: only rarely and under the most favourable conditions can the mat reach up to 3–4 cm in diameter.

As early as 2 June, swelling, glossy buds were observed, bright crimson in colour and 1–1.5 mm in diameter, on thawed out patches near Zaliv Dezhneva. Individual buds had started to open. On 12 June at the same site, folded leaves, *c.* 2 mm long, were observed emerging from a part of the leaf buds, although many were not yet open and were still violet coloured. On 15 June many leaves had grown to 8 mm in length and the longest were beginning to unfold. Greening tufts of *S. hyperborea* which had become free from snow at an early date were first observed at this site on 22 June, when all the leaves, growing out from the leaf buds, had unfolded. On 1 July the first half-open flower buds were noticed.

The first sessile flowers appeared on 11 July and mass flowering took place between 20 and 25 July. On 1 August the flowers started to wilt, the dimensions of the ovaries increased and the peduncles began to elongate. By 6 August the peduncles had reached various lengths: below some flowers they were short and the lower part of the sepals remained hidden by the leaves, below others they were long and the flowers protruded a few mm above the tufts. On the same date, autumn coloration started to appear: the mats took on a purplish or vermilion-reddish colour; the peduncles, elevated above them, continued to elongate up to 20 August. The rest of the flowers remained, as before, almost sessile. By 30 August all the leaves had dried and become brown, and the non-fruiting stems had also turned brown. The seeds were almost exclusively retained by those flowers which were nearly sessile; their dark violet, brownish and glossy capsules did not open. The capsules were 4 mm broad below the horned stigma and reached *c.* 2.5 mm beyond the sepals. They were buried beneath the snow in that condition.

Thus, in the case of *S. hyperborea*, the development of two kinds of peduncles can be observed. Some grow out and are 12–15 mm above the surface of the leaves when the capsule starts to develop. The rest of the flowers remain sessile initially so that only their petals protrude above the rosette leaves; following anthesis, their peduncles grow a little longer, usually up to 3 mm, and the flowers become slightly elevated above the mat but the lower portion of the sepals remain covered by the leaves.

No capsules were seen to open during the autumn but in the spring, 2–3 weeks after the mats had become free from snow, many of the capsules from the previous year were seen opening. These opening capsules had been formed almost everywhere by the sessile or nearly sessile flowers, but the flowers on the long peduncles did not produce ripe capsules as a rule; they were covered by the snow in that condition during the autumn. Apparently the advantage of the small extra amount of heat which the sessile or nearly sessile flowers had received in comparison with that received by those on elongated peduncles plays an important role.

Specimens of *S. hyperborea* are able to flower during the short growing period at sites where the snow disappears only by 15 July. At such sites I frequently encountered individuals in open aggregations which were very small in August. Where the snow disappeared even later, they rarely occurred and then only in the form of tiny, sterile mats.

Saxifraga caespitosa L. is a cushion-forming plant with an arctic–alpine circumpolar area of distribution. It is rarely found on Zemlya Aleksandra, and then only as single individuals, mainly at Zaliv Dezhneva, where thawed-out patches appear early. It is met with only in places where the snow melts between the end of May up to as late as 22 June. It grows in the form of dense cushions, usually 3–3.5 cm in diameter, rarely 4–4.5 cm and very rarely up to 5 cm in diameter.

As early as 2 June, leaves which had survived the winter were observed growing on specimens of *S. caespitosa* in habitats near Zaliv Dezhneva that became free from snow at an early date; the leaves had survived the winter in a green state but were deep inside the rosette, completely hidden below a mass of dead, brown leaves from the preceding year. The cushions had turned completely green by 20 June when these leaves had emerged from below the material from years past. At this time also the overwintering buds began to be noticed, still closed, sessile and red-coloured, deep inside the leaf rosette ('the tussock'). By 6 July the buds were swelling, their colour lilac to dark red.

The first flowers were observed on 9 July, sessile and buried inside the cushion. Mass flowering began on 14 July and on 25 July it reached its peak. On 23 July the formation of capsules was noticed in those flowers which had bloomed first, i.e., those on the south-facing side of the cushion. On 15 August the flowering was still going on (the flowers remained, as before, sessile) but by then the post-anthesis stage and fruit-setting were the predominant phases. By 20 August all the flowers in the area around Zaliv Dezhneva were finished and a stage of mass fruiting had started. The capsules formed did not protrude above the leaves of the cushion. It was not possible to see whether any of the capsules had opened.

S. caespitosa is met with, although rarely, in other areas of the island where the snow disappears at a relatively early date, but here its development is limited to the vegetative growth and bud-setting stages; a few flowers began to appear only during the autumn.

My observations of a cushion of *S. caespitosa* growing near Zaliv Dezhneva in habitats where the snow melted at the end of May were as follows:

On 2 June the interior, overwintering leaves were hidden inside the rosette by the external, dead brown leaves although there were already signs of a renewal of the vegetative stage: the buds at the base of the rosette, where it was lying pressed to the ground, were opening and it was possible to notice the green, interior leaves with their reddish tips. On 12 June all the buds on the rosette, even those in the upper parts of the cushion, were found in a half-open condition. By 15 June the leaves with bright crimson tips had elongated inside the rosette and were the same size as the dead leaves from previous years. On 19 June sessile, closed, red-coloured flower buds could be seen at the bottom of the rosette, which were most open. On 1 July these buds were still closed, sessile and red. On 6 July the rosette appeared bright green and red: the living leaves, lending this colour to the cushion, completely covered the dead material from the previous year. The flower buds were still closed, lilac to dark red, and were buried inside the cushion 1–2 mm below the tips of the leaves. On 9 July fourteen flowers on the south-facing side of the cushion had started to open: they were sessile and 1–2 mm below the tips of the leaves. On 14 July twenty-eight flowers were completely open in the same part of the cushion: they had elongated somewhat, the edges of their petals were level with the tips of the leaves in the cushion. On 23 July there were sixty flowers all over the cushion and most of the flowers on its south side had wilted: their petals were slightly faded and brownish and the ovaries were swelling. The latter were green, the stigmas rose-coloured. The tips of the petals reached above those of the leaves in the cushion. On the north-facing side of the cushion it could be seen that among the flowers that were opening up there were many which were still only half open. On 7 August fifteen flowers were still in bloom, the rest were finished; the ovaries were full and swelling. Even at the post-anthesis stage, the ovaries did not reach above the leaves in the rosette. On 18 August capsules were observed to be forming: they were sessile and red- or lilac-coloured, as before. No further observations concerning the opening of the capsules could be made that year.

Saxifraga oppositifolia L. is a circumpolar, arctic–alpine species. It occurs as caulescent, tap-rooted and polycarpic herbaceous specimens (Polozova, 1978). Gorodkov (1958*b* and others) and Korotkevich (1958) mistakenly categorized it as a dwarf shrub in spite of the fact that this plant, according to Warming (1912) is herbaceous. It is rarely met with on Zemlya Aleksandra but it can occasionally be found on substrates where the fine soil is mixed with stones, and under conditions of an optimal snow regime.

In the opinion of the translator, because of its somewhat atypical morphology, ecology and phenology it is very likely that this is the same high-arctic tetraploid taxon that is found on Spitsbergen, i.e., *S. pulvinata* Small (syn. *S. oppositifolia* ssp. *Smalliana*) (see Flovik, 1940; Hedberg, 1967; Zhukova, 1967; Johnson and Packer, 1968; Petrovsky and Zhukova, 1981).

The best developed specimen of *S. oppositifolia* was observed on 14 August in the eastern part of the Zaliv Dezhneva area, where it was growing on medallions inside a stone net and at an elevated level on a pebbly-sandy marine terrace on sites with a south-facing exposure. It was growing in the form of a well-developed mat, in which the dead, dark-grey shoots were almost completely hidden by fresh leaves and new, living shoots, part of which had grown well during that season. On this date, *S. oppositifolia* was still flowering relatively abundantly but the majority of the flowers had wilted and showed ripening capsules. In a number of cases it was possible to see open capsules: these capsules had been established the previous season.

In the most westerly part of the shore along Zaliv Dezhneva and also in relatively elevated plakor habitats in the northern part of the island, *S. oppositifolia* was found at only a small number of sites (Table 7) and was considerably less well developed: dead shoots predominated, giving the mats a blackish colour and only a few flowers were noticed in July. The first flowers had wilted by 7 July. No ripe capsules were found.

The more satisfactory development of *S. oppositifolia* in the eastern part of the shore around Zaliv Dezhneva depended on the fact that this part of the shore has more favourable relief characteristics and is better protected from the northerly winds.

Thus, on Zemlya Aleksandra, *S. oppositifolia* does not behave like an early spring-flowering plant with rapidly advancing phenological stages, such as is characteristic of it in the arctic tundra. On the contrary, here its flowers appear late, considerably later than in such angiosperms as *Cochlearia groenlandica*, *Draba oblongata*, *D. pauciflora* and *D. macrocarpa*. Anthesis extends into the autumn (a post-anthesis stage occurs simultaneously with the appearance of fresh flowers) and ripe capsules form only under the most favourable conditions.

The observations on the seasonal development of the flowering plants on Zemlya Aleksandra demonstrate that in the extremely severe conditions prevailing there the species appear to have undergone a curious adaptation, ensuring not only their survival but also their propagation and, thus, the continued existence on the island of differently aged populations. Immature specimens of many species were frequently found growing together with older individuals of various ages.

Concerning the kinds of adaptations ensuring the survival and propagation of the flowering plants, it is first and foremost necessary to point out the predominance of life forms, such as cushions and mats, both compact and loose. Matveyeva (1979) emphasized the predominance of such growth forms also in the area of Mys Chelyuskin. This is expressed not only in the distribution of the species, for which such life forms are in general characteristic (e.g., *Phippsia algida*, *Poa abbreviata*, *Cerastium arcticum*, *C. regelii* ssp. *caespitosa*, *Papaver polare*, *Draba oblongata*, *D. pauciflora*, *D. macrocarpa*, *D. subcapitata* and *Saxifraga caespitosa*) but also by the fact that some species have assumed a cushion- or tussock-like shape that is not their characteristic habit in the tundra zone. In the arctic and subarctic tundras *C. regelii* (in this case *C. regelii* ssp. *regelii*) is represented by prostrate forms (Polozova, 1978) while on Zemlya Aleksandra the high-arctic taxon (ssp. *caespitosa*) forms dense cushions. Such species as *Saxifraga cernua*, *S. nivalis* and *S. tenuis* grow in the tundra zone characteristically in the form of solitary stems but on Zemlya Aleksandra they form mats, just like *S. hyperborea*. Matveyeva (1979) writes about the growth of *S. cernua* in the form of mats also on Mys Chelyuskin.

The cushion-like growth form contributes to the conservation of heat (an analogy can be drawn with the fur of animals) both in respect to solar energy reaching the cushion and the energy conducted from the ground, which is always much warmer than the air (also as a consequence of the effect of solar radiation) (see Fig. 51).

Because the ground is warmer than the air, a rosette plant like *Cochlearia groenlandica*, which does not usually form a mat, grows in the form of very small rosettes, the leaves of which are pressed flat to the ground. The flower buds of *Papaver polare* are also lowered to the ground and, especially, to the stones (which during daytime hours are warmer than the fine soil; see p. 24) when the peduncles are so elongated that the buds can be seen above the cushion. Because the ground is warmer than the air, renewed vegetative growth in the spring begins among the lower shoots of the clumps, i.e., those touching the ground. Thus, during my year of observations, the development of regenerating shoots of *Poa abbreviata* occurred mainly in the lower part of the turf and only later on during its development did culms appear, in a number of cases growing horizontally and pressed to some stones on the ground.

The presence of sessile flowers is an adaptation which assures a favourable development of the buds, flowers and ripening seeds. Sessile flowers are found on the majority of the cushion-forming species: at the start of their development the flowers are buried within the cushion formed by the basal leaves, and only when the seeds are ripe do they begin to elongate (because of the growth of the peduncles) so that ripening siliquas or capsules become just slightly raised above the cushion, in which the base of the sepals remain either covered by the leaf tips or level with them. Sessile flowers are found on those species which in the tundra area have well-developed peduncles. On Zemlya Aleksandra this is the case for all the species of *Draba* as well as for *Cerastium arcticum* and *Saxifraga caespitosa* and others. Following the ripening of the seeds, the siliquas or the capsules are raised above the mat (or cushion) thanks to the elongation of the peduncles, as a rule to just above the surface of it, i.e., just enough so that they will not open inside the cushion but outside of it so that the seeds can be caught by the winds. The advantage of sessile flowers is that they are sheltered from the winds and the snow that frequently falls during the summer, and also utilization of the small amount of extra heat from the cushion is optimized; this can easily be seen in the case of e.g.

Saxifraga hyperborea. In this species some of the flowers are sessile, just like the ripening capsules: even when these protrude above the cushion, their bases remain covered by the tips of the rosette leaves. Other flowers have peduncles which increase in length and lift the flowers by some 12–15 mm above the cushion at a stage when they have wilted. As a rule these flowers do not develop ripe capsules but dry up during the autumn, when they become covered by snow; simultaneously the sessile flowers form capsules, which open when ripe during the following season (see p. 188).

Papaver polare does not have any sessile flowers. In fact, long peduncles develop, just as in the tundra, but the flowers are not raised above the surface of the cushion. Rather, they are pressed against it, and thanks to the capacity of the peduncles to react thermotropically, they direct the buds with striking accuracy towards the surface of the cushion so that they are always found not far above the level of the leaf tips, i.e., not fully raised above it, but neither completely buried inside the cushion. However, should the peduncles carry the buds outside the limits of the cushion, they direct the buds towards the ground and, mainly, towards adjacent stones (for more details, see p. 170).

The ripening of seeds during the following season seems also to be an important adaptation. Siliquas opening during the year when formed were observed only in the case of *Cochlearia groenlandica*. Of the species developing their generative organs most rapidly, i.e., *Draba oblongata, D. pauciflora* and *D. macrocarpa*, it was observed that some of the siliquas opened the same year but that the rest of them opened only during the following season. The opening of siliquas of *Draba* spp. and capsules of *Cerastium arcticum, Saxifraga hyperborea* and *Papaver polare* was observed taking place at various dates during the year following their development after overwintering and ripening in the course of the second vegetative season.

Still another kind of adaptation appears to occur in the case of *Phippsia algida* and *Saxifraga cernua*: this takes the form of biennial, and in the case of *S. cernua* also triennial, regenerative shoots which survive the winter(s) alive and emerge from the snow capable of functioning during the season(s) following and die off only when, in the case of *P. algida*, the seeds have ripened or, in the case of *S. cernua*, new proliferating shoots have formed (for detailed descriptions of these phenomena, see pp. 161 and 186).

Not only the small dimensions of the plants but also the slow development of regenerating shoots contribute to the survival of individuals and the ripening of the seeds: plants may be thought of as using the minimal resources which the severe nature of the arctic polar deserts offers them as economically as possible. On days when snow falls and the temperature drops below the freezing point (Fig. 51) the development of individuals is halted but the flowers are not killed by the frost, not even on the poppies. Thereafter every – even the most unimportant – rise in temperature is utilized by the plants for continuing their seasonal development.

The change in seasonal aspect that has been described for the tundra associations (e.g. Shamurin, 1966*b*) is not characteristic of the arctic polar deserts. This difference in seasonal aspect depends first and foremost on the fact that flowering plants occupy such small areas and that each specimen is very tiny and does not easily catch the eye. It is usually necessary to search for them in order to be able to produce a geobotanical description. Because of this, not even the simple aspect of 'turning green' is obvious there.

The lack of seasonal aspects in the arctic polar deserts depends also on the predominance of sessile flowers, which are hidden within the cushions during both the budding and the flowering stages. *Papaver polare* represents an exception with its large and brightly coloured flowers. However, habitats with a relative abundance of flowering poppies are met with only rarely, just in a few aggregations belonging to the type B nanocomplex and *comitium* no. 3 (Tables 8 and 12): but such sites have few poppies (the estimated cover does not exceed 4% even in the best cases). In addition, the poppy flowers are raised above the ground by not more than 3 cm, and only during the warm hours, which occur so rarely, are the majority of the peduncles inclined so that the flowers either just touch the ground or are slightly raised above it. This reduces the opportunities for observing any seasonal aspects of the vegetation.

The aggregation serving as the standard for the zonal vegetation of the northern belt of the polar deserts in the Barents province (i.e., nanocomplex type B (see Table 9)), is characteristic of the exceptional monotony of the appearance of the surface, which does not vary in the course of the entire growing period (Aleksandrova, 1971). In the southern belt of the arctic polar deserts more colourful extrazonal associations occur (Korotkevich, 1958; Safronova, 1979, 1981*a*).

6

Conclusions

When the data set out above are analysed it becomes possible to characterize the distinguishing traits of the area, covered by the arctic polar deserts, as those of a special geobotanical region in a manner more exact than was done in my previous works (Aleksandrova, 1969, 1977a, 1980). These include the types of vegetation, in both plakor and non-plakor habitats (which especially strongly distinguish the vegetation of the polar deserts from that of the tundra region) as well as the bio-morphological specialization of the plants, the structure of the flora and the level of activity of the species.

The vegetation on plakor habitats (zonal aggregations)

The plant aggregations developed on relatively well-drained plakor habitats with loamy soil are of decisive importance for the recognition of geobotanical areas according to the concepts used in this book.

These concepts, which were based on ideas set forth by V. V. Doku-chayev, N. M. Sibirtsev and I. K. Pachoskiy, were introduced into Soviet geobotany by Lavrenko (1947) as a method for establishing geobotanical areas. They received widespread application on many published maps, where the distinction into regional subdivisions was revealed by clearly differentiated floristic characteristics, by full correlation of natural associations with natural conditions and indices of economical utilization of the plant cover. In addition to the zonal associations, which had the heaviest weighting in the determination of these areas, all the variously formed non-plakor types of vegetation were included (Lavrenko, 1947).

On the periglacial territory of high-latitudes in the Arctic, the plakor aggregations have characteristics which enable them to be included in the basic polar desert type of vegetation as distinct from that of the tundras (Aleksandrova, 1969, 1977a; Matveyeva and Chernov, 1976; Matveyeva, 1979): dwarf shrubs are absent from these aggregations, while they appear as constant and characteristic components of the zonal tundra associations. Other flowering plants are few in number and poorly developed: their dimensions are sharply reduced and they have root systems which do not intermingle (Bell and Bliss, 1978, have also discussed the poor development of root systems in the Canadian polar deserts on King Christian Island) while in the tundra phytocoenotic associations with root systems which do intermingle are prevalent (see Aleksandrova, 1962, 1977a, etc.). In the polar deserts cryptogamic plants predominate in terms of both cover and phytomass, while in the tundra phytocoenoses the bulk of the phytomass is made up of flowering plants (Aleksandrova, 1977a). In the polar deserts the angiosperms completely lack any kind of edaphic function; the special biology and morphology of the angiosperms and of the cryptogams are characteristic: the arctic–alpine crustose lichens are of the greatest importance for the structure of the plant aggregations, forming characteristic synusia within the type of vegetation in question.

It is necessary, however, to emphasize that plakor habitats with loamy soils and a corresponding development of plant aggregations on them are very rarely found within the arctic polar deserts. This is due, first and foremost, to the absolute predominance there of stony habitats (concerning the importance thereof, see p. 14 above) with a saxicolous vegetation and, secondarily, to the fact that the aggregations on plakor habitats, of which there are few, are found only in areas of narrow ecological niches with an optimal snow regime. These are habitats where the snow never reaches a depth of more than 25 cm, while in the Arctic and the northern subarctic tundras (where the total sum of the temperatures is higher and the snow melts off sooner) the depth of snow on plakor habitats with zonal associations occupying a considerably larger surface area can be as great as *c.* 40 cm (Aleksandrova, 1961, 1962; Matveyeva *et al.*, 1973). Thus, the vegetation on plakor habitats covers a considerably smaller surface area within the polar deserts than on the plains belonging to the tundra zone.

The vegetation on non-plakor habitats

As demonstrated on the preceding pages, in the arctic polar deserts even a minor deficiency in the amount of snow will lead to the destruction of nanocomplexes with a continuous plant cover and to the formation of open aggregations (semi-aggregations and aggregations). In the polar deserts the surface area occupied by open aggregations in non-plakor habitats exceeds by at least the power of ten the surface area occupied by associations with a closed type of plant cover (Table 22). In addition, much of the land mass is occupied by sites which completely lack any plants. On Zemlya Aleksandra this amounts to 25% of the surface.

The fact that the predominant portion of the territory is occupied by open aggregations sharply distinguishes the vegetation within the area of the arctic polar deserts from that within the tundra zone. In the latter, closed phytocoenoses predominate, but open associations, although sometimes met with, are rare and found only under extreme conditions. They may be encountered in some areas in the form of non-continuous strips or cushions, forming associations in habitats from which the snow is completely blown away, and sometimes, although only in the northern part of the tundra zone, also where snow cover persists until a late date, i.e., into the latter part of the summer. In addition, these open associations occur under fairly unfavourable edaphic conditions, e.g., within cryo-active boulder fields, etc. On the whole, they occupy only a small percentage of the territory within the tundra zone. It is also important that in general there are no sites in tundra areas which are completely lacking in vegetation: even where the snow disappears late there are always habitats where a certain amount of plant cover develops. The vegetation of nival, meadow-like associations and snow bed associations in the tundra zone has been described by many authors. In the arctic tundra of Novaya Zemlya there are depressions, occupied by mires, where a deep snow cover piles up. These become free from snow about two months later than the areas which thaw out first (Aleksandrova, 1956). However, even there a beautifully developed mire vegetation can be found. Also, at those locations under steep slopes where the snow lasts until the middle of August, some individuals of *Saxifraga oppositifolia*, *Cerastium regelii* and other species can be found: sites where plants are entirely absent have never been observed within the limits of the plains at low altitude. Such sites begin, however, to appear on the South Island of Novaya Zemlya, in the mountains at the transfer into the belt of alpine polar deserts.

Thus, the vegetation on non-plakor habitats very definitely distinguishes the region of the polar deserts from that of the tundra zone. The absence of peat bogs is also characteristic of the polar deserts: there are only so-called small bogs on mineral soil without peat: these occur in the form of a thin sward of moisture-loving mosses percolated by water, among which may be seen some flowering plants, although only in small quantities.

The specificity of the biology and the morphology of the plants

The flowering plants of the arctic polar desert are distinguished not only by their scarcity but also by the low number of species and by a number of biological and morphological characteristics related to their adaptation to the extreme conditions. The distinct lack of heat appears to be the most important factor (see pp. 155–95). Among the number of adaptive adjustments of the cryptogams, the particular morphology of the liverworts is especially significant in the polar deserts.

The structure of the flora

The data thus far presented allow a detailed analysis of the peculiar structural features of the flora within the region of the arctic polar deserts. The particular characteristics of this flora are evident first and foremost by the fact that the lichens are greatest in number, the mosses hold second place and the vascular plants come only in third place. The latter are represented exclusively by flowering plants (no vascular crypto-gams are found). Thus, on Zemlya Frantsa Iosifa, 115 species of lichens, 102 species of mosses, 57 species of flowering plants and 55 species of liverworts are known (according to A. L. Zhukova and R. N. Shlyakov, personal communication); on Mys Chelyuskin, 136 species and subspecies of lichens (Piyn, 1979), 74 species and subspecies of mosses (Blagodats-kikh *et al.* 1979), 59 taxa of flowering plants (Safronova, 1979) and 25 taxa of liverworts (Blagodatskikh *et al.*, 1979) have been observed.

There are no *Sphagnum* spp. known for the moss flora of the arctic polar deserts, and among the lichens many are crustose.

For comparison with the foregoing figures consider the composition of the flora at the Tareya Research Station (in western Taymyr), situated within the northern belt of the subarctic tundra: there the vascular plants are most abundant, with 239 taxa (Polozova and Tikhomirov, 1971), the mosses occupy second place, with 162 taxa, of which 17 belong to *Sphagnum* (Blagodatskikh, 1973), and the lichens third place, with 117 taxa (Piyn and Trass, 1971). In fourth place come the liverworts, with 46 species (Zhukova, 1973*a*).

The composition of the soil microflora in the polar deserts is also peculiar. The blue-green algae predominate among the soil algae and develop both in the soil and on its surface, while in the tundra the green algae hold first place (Novichkova-Ivanova, 1972).

The flowering plant flora has also peculiar characteristics. First of all, it is extremely poor in numbers of taxa (Table 6). The areas which lie within the southern belt of the arctic polar deserts have a very small number of species compared to the tundra region: on Severnaya Zemlya, 75; on the Nordaustlandet of Spitsbergen, 67. They are situated north of the so-called 'isotaxon 80' (Beschel, 1969). In the northern belt of the polar deserts, the poverty of this flora is even more extreme as on Zemlya Aleksandra there are only 24 species of flowering plants and on the northern islands of the Severnaya Zemlya archipelago there are not more than 17 species.

The species composition of the flowering plant flora

The leading families of flowering plants found in the region of the arctic polar deserts are: the Gramineae, with 22 species and subspecies; the Cruciferae, with 17 taxa, the Caryophyllaceae with 14 taxa and the Saxifragaceae with 13; the leading genera are: *Saxifraga* with 12 species and subspecies and *Draba* with 11, followed by *Poa* with 7, *Stellaria* with 5 and *Ranunculus* with 4. The almost complete absence of species belonging to the Cyperaceae is one of the significant differences from the tundra flora, while the Juncaceae is represented by 3 species, i.e., *Juncus biglumis*, *Luzula confusa* and *L. nivalis*, which act as constant members of the arctic polar desert flora and are met with in all the areas so far investigated.

In the tundra flora, the Cyperaceae is one of the leading families, and the entire complex of sedges and cotton grasses is found in the most important tundra associations and in those of the tundra mires. In the flora of the arctic polar deserts, however, the representatives of this family are almost entirely absent: the four species of Cyperaceae found within the area of the Soviet arctic polar desert (see Table 6) occur only sporadically within the southern belt of the polar deserts and are met with there extremely rarely. Thus, e.g., *Carex ensifolia* ssp. *arctisibirica* has been found only once on Mys Vatutina on the Zhiloy peninsula (on Ostrov Oktyabr'skoy Revolyutsii, Severnaya Zemlya), where, as stated above (p. 138), there are local temperature anomalies. It was discovered there in 1958 by I. S. Sey, who found only a single tussock of this sedge (see Safronova, 1981*b*).

The gamopetalous species also appear to be alien to the arctic polar desert flora: individual representatives of the Primulaceae, Boraginaceae, Scrophulariaceae and Compositae reach only into the southern belt of the region and are met with there only in extrazonal habitats. In the northern belt of the polar deserts the list of species of flowering plants is (when following Engler's system) 'cut off' behind the Rosaceae, and, as stated above, the flora is very limited and impoverished.

Such peculiarities in the structure of the flowering plant flora supplements the number of differences between the regions of the polar deserts and those of the tundra.

As far as the geo-elements of the flowering plant flora are concerned, arctic (including high-arctic) and arctic–alpine species predominate. These two groups of species play an equal part in the composition of the angiosperm flora within the limits of the Soviet arctic polar deserts. Only three of the species, which have been reported only recently, have a more southerly type of area of distribution: *Chrysosplenium alterniflorum* (an arctic–boreal species), *Artemisia borealis* (a hyparctic–montane species) and *Nardosmia frigida* (an arctic–boreal species). These have been found in the southern belt of the polar deserts in extrazonal habitats, i.e., in extrazonal associations of tundra type. In aggregations belonging to a polar desert type of vegetation only arctic (including high-arctic) and arctic–alpine species take part.

Various maps show the composition of the geo-elements within the cryptogamic flora in the polar deserts. At the same time as flowering plant species with arctic and arctic–alpine areas of distribution are met with in equal amounts, the cryptogams which form the basis of the polar desert aggregations are represented almost exclusively by arctic–alpine species. The relatively large number of mosses and lichens with arctic–alpine bipolar areas of distribution also constitute a major difference from the tundra region. Among the flowering plants species with bipolar distribution patterns are non-existent: there is only one bipolar genus, i.e., *Deschampsia*, to which the Antarctic *D. antarctica* belongs. The following mosses and lichens occur within both the Arctic and the Antarctic: *Drepanocladus uncinatus, Distichium capillaceum, Pohlia cruda, P. nutans, Polytrichum alpestre, P. alpinum, Rhacomitrium lanuginosum, Usnea sulphurea, Alectoria nigricans, A. pubescens, A. minuscula, Sphaerophorus globosus, Rhizocarpon geographicum, Lecanora polytropa, Rinodina turfacea, Psoroma hypnorum, Ochrolechia frigida, Buellia coniops, B. punctata* and some others.

Oksner (1944) studied the origin of the bipolar distribution of lichens. He states (p. 255) that: 'In the arctic–alpine–antarctic element the most dominant group is formed by species the generic roots of which have their origin in the flora of mountain regions of the temperate parts of the Holarctic. Among those species we can find *Toninia candida, Haematomma ventosum, Lecanora epibryon, Cladonia elongata*, and so on. Antarctic centres play a considerably lesser part in the formation of bipolar species (e.g., *Psorama hypnorum* or *Neuropogon sulphureus*).' As far as an arctic origin of the bipolar species is concerned, it is according to the opinion of Oksner 'possible to suspect' that this is valid for only two species of *Lecidea*. In addition, this author writes (Oksner, 1944, p. 256): 'Many have attempted to refer the origin of bipolar distribution patterns to the chance dispersal of diaspores. Only the causal-historic methods, first applied by Charles Darwin and J. Hooker, furnished a solution to this problem in general terms. None of the schemes suggesting a pathway for bipolar species over the Panama isthmus can stand confirmation during an analysis of migratory routes. The dispersal of bipolar species is feasible only via the Venezuelan–Colombian mountain chain and the Antillean fold area, which is presently submerged (i.e. the so-called *via Antillana*).' Some of the arctic–alpine mosses and lichens distributed within the arctic polar deserts occur in more than one area, e.g., *Rhacomitrium lanuginosum* is an almost cosmopolitan species which is widely distributed within the Arctic. *Sphaerophorus globosus* and a few other species have a similar type of distribution.

There are few species of mosses within the polar deserts which have an exclusively arctic distribution. Among those which do we find *Polytrichum fragile, Psilopilum cavifolium, Seligeria polaris, Bryum nitidulum, B. teres, Distichium hagenii, Drepanocladus latifolius, Campylium zemliae* and some other species, *Hylocomium alaskanum (H. splendens* var. *alaskanum*) and *Tomenthypnum nitens* var. *involutum* (the latter two are races of hyparctic species).

There are also only a few species of lichens with an exclusively arctic area of distribution. Thus, e.g., from the list of the 125 species of crustose lichens published for Mys Chelyuskin by Piyn (1979), only 12 species are mentioned as having an arctic distribution and, in addition, they occur within that area only as 'rare' or 'very rare' taxa (they are *Leciophysma finmarkicum, Parmeliella arctophila, Arctomia delicatula, A. interfixa, Bilimbia subfuscula, Lopadium coralloideum, Hypogymnia oroarctica, Cetraria elenkinii, C. simmonsii* var. *intermedia, Siphula ceratites, Pertusaria subdactylina* and *Placodium subfruticulosum*). Thus, species with arctic–alpine areas of distribution (113 out of 125) predominate. The most active species of crustose lichens on Mys Chelyuskin belong, as stated by Piyn (1979), to species with arctic–alpine distribution areas: *Psoroma hypnorum, Pannaria pezizoides, Cetraria cucullata, C. delisei, C. islandica* var. *polaris, Dactylina arctica, Parmelia omphalodes, Thamnolia subuliformis, Stereocaulon rivulorum, Ochrolechia frigida* and *Lepraria arctica* (all listed in systematic order).

It is thus possible to confirm that among the spore-producing components of the plant associations of the arctic polar deserts, absolute predominance is exerted by arctic–alpine species. It is especially necessary to emphasize that the characteristic synusiae of a polar desert type in this vegetation are formed by crustose lichens with an arctic–alpine area of distribution area (*Ochrolechia frigida, Toninia lobulata, Collema ceraniscum* and *Pertusaria glomerata*, etc.). They form the basic crust on the ground surface and on the remnants of plants, where they are associated with *Rinodina turfacea, Pannaria pezizoides, Psoroma hypnorum, Ochrolechia grimmiae, O. gonatodes* and other species, which also have arctic–alpine areas of distribution. In the crust of crustose lichens which plays a major role in the composition of the polar desert aggregations, it is not unusual to find some inclusions of arctic–alpine species of mosses in the form of various small shoots of *Polytrichum alpinum, Pohlia cruda* and sometimes also of *Ditrichum flexicaule*, the only one among the mosses which appears to be an especially typical component of the plant aggregations of the polar deserts.

The absolute predominance of the arctic–alpine species allows the suggestion that the arctic polar desert type of vegetation, in its form of a definite floro-coenotype, may already have begun to be formed during the time of the Late Pleistocene era of falling temperatures and extreme conditions in high-alpine areas near the snow line, where, in the process of adaptation to present living conditions, the moss–lichen synusiae which are now characteristic of the polar desert type of vegetation, were formed. It is very likely that the polar desert synusiae were finally established during the Glacial Period in arctic and subarctic areas which were free from ice cover and in various alpine refugia. In the territories at present occupied by the arctic polar deserts, refugia were entirely absent during the Glacial Period, and those synusiae, of which we speak here may have started their adaptation in areas which only later became free of ice, i.e., when periglacial land started to emerge from below the ice at high latitudes. Then also species of flowering plants might have started to invade and take part, although not to any major extent, in the composition of the plant aggregations of the polar deserts.

The territory of the arctic polar deserts was distinguished as a special region of major importance by Passarge (1920) under the epithet of 'cold deserts', by Berg (1928, 1936), who called it the 'glacial zone' and by Gorodkov (1935), who designated it as the 'nival zone' and who distinguished the subzones of 'arctic deserts' and 'arctic desert-glacial areas'. Later, Gorodkov (1958*a*) widened his concept of 'arctic deserts' and determined its limit towards the south; his limits have been subjected to criticism (see Aleksandrova, 1977*a*).

Leskov (1947) called the region in which the arctic polar deserts are situated a 'high-arctic nival region'; Korotkevich (1967, 1972, etc.) named it an 'arctic zone of polar deserts'; Aleksandrova (1969, 1971, 1977*a*, etc.) gave it the epithet a 'region of arctic polar deserts'; Matveyeva (in Matveyeva and Chernov, 1976; Matveyeva, 1979) called it a 'zone of polar deserts'; Tedrow (1970; 1972) referred to this territory as a 'polar desert zone'; Bliss (1977) distinguished polar deserts and semi-deserts in the northern portion of the Canadian Arctic Archipelago, just like Tedrow, including in them areas where considerable surfaces are occupied by alpine polar deserts (as a result of the presence of vertical belts) established within the limits of the subzone of arctic tundra.

Recently, a scheme for the botanical-geographical division of the Arctic has received wide attention, i.e., that drawn up by Yurtsev (1966) and based on a method of studying specific floras and, more recently, also 'partial floras', which specify ecological niches for the floristic complexes revealed (Yurtsev, 1975; Yurtsev and Semkin, 1980, etc.).

Yurtsev uses a number of characteristics for the distinction of botanical-geographical regions, but according to his concept the floristic composition of the vascular plants appears to be the diagnostic characteristic with the greatest weight. This can be established by means of a study of specific floras, the components regarded as having the greatest importance being 'aggressive' species: 'As botanical-geographical criteria of importance for zonal division, I employ data concerning changes from the south towards the north in the composition of aggressive (i.e., successful) species of plants, which compose the basis of the plant cover' (Yurtsev *et al.*, 1978, p. 20); and 'precisely that change in relationship between aggressive species, which represents various latitudinal groups and, in general, reflects the behaviour of the plants in respect to the total amount and/or intensity of summer heat, represents the most universal criterion for a zonal distinction of a vegetation cover; this criterion works within very different longitudinal sectors and makes it feasible to homologize comparable latitudinal divisions within their limits' (Yurtsev *et al.*, 1978, p. 21). In his papers Yurtsev's conclusions are drawn on the basis of analyses of the floras of vascular plants.

Recently, the school of B. A. Yurtsev has paid attention also to the study of specific floras among the cryptogams (Andreyev, 1981, etc.), although the results of these investigations are not yet, like other works concerning the study of specific cryptogam floras (Afonina, 1974; Abramov *et al.*, 1975, etc.), comparable to the botanical-geographical zonation drawn up on the basis of analyses of the floras of vascular plants.

The principles used for a phytogeographical zonation include a number of differences from those used for geobotanical zonation, such as I have presented in this book, where the major emphasis is placed on the phytocoenological specificity of the plant aggregations, and in particular their structure and their entire floristic composition, including all the groups of the cryptogamic components. This method also allows the distinction into ranks of separate regional subdivisions. In the present case it is important to stress the fact that, according to Yurtsev, the polar desert territory cannot be distinguished at a level of higher rank: he includes it in the tundra zone as 'a glaciated variant of the high-arctic tundra'.

The arguments set forth by Yurtsev lead basically to the fact that the differences in the composition of the vascular flora at high latitudes (on which the main attention of this author is fixed) '... are in principle of a quantitative character (reduction in size and variation, a slowing down of the speed of processes). However, even if qualitative, they are definitely negative, leading to a further elimination of a whole row of taxonomic groups and, as a consequence thereof, to a redistribution of the groups remaining within the composition of the plant cover and represented in particular by the increase in the relative role of the cryptogams' (Yurtsev *et al.*, 1978, pp. 18-19). He does not give any significance to the structural character of the plant aggregations, which is reflected in the basic role (for the entire distinction into areas) of the plakor aggregations. He also puts forth an argument concerning life forms, but that seems to be based on a misunderstanding (see Aleksandrova, 1979, p. 1719, footnote).

Thus, Yurtsev (Yurtsev *et al.*, 1978, p. 19) states that in the composition of 'the periglacial variants of the high-arctic tundra', dwarf shrubs are actually met with. For this argument he refers to a short statement in my lecture on the vegetation of Zemlya Aleksandra (in the archipelago of Zemlya Frantsa-Iosifa (*Bot. Zhurn.* **45** (12), p. 1821, 1960)) where, according to Yurtsev, I should have mentioned the distribution of dwarf shrubs in the composition of polar desert aggregations. However, Yurtsev was mistaken; I did not mention any dwarf shrubs because they are entirely lacking on Zemlya Aleksandra. In my papers I have repeatedly stated that one of the characteristics of the polar deserts is the complete absence of any dwarf shrubs.

Therefore the botanical-geographical division into areas according to Yurtsev, constructed by him on different, mainly floristic principles and of an independent nature, is not compatible with my own division into areas, which is determined by other kinds of criteria.

Aggressiveness of the plant species in the polar deserts

It is of interest to apply the criterion of aggressiveness as defined by Yurtsev (1968, pp. 156-8) in a comparison between the vegetation of the arctic tundra and that of the arctic polar desert territories. It is quite possible to convince oneself that in this respect there are also major differences between these two territories.

In the arctic tundra region there are usually species with 'first degree aggressiveness' (i.e., particularly aggressive species which occur in abundance and are eurytopic). To these belong *Salix polaris* in some areas of the arctic tundra and *Eriophorum vaginatum* in Chukotka, etc. In the tundra territory many species, both flowering and spore-bearing, are characterized by 'second degree aggressiveness' (especially the high-arctic species); these occur abundantly and are either rather widely distributed or few in number but still eurytopic.

During invasion into the arctic polar deserts, the general aggressiveness of a species falls off sharply, owing to the fact that in large areas in general no plants at all are met with. Because of this, the special characteristic of the polar deserts appears primarily to be the complete lack of eurytopic species, which excludes the presence of species with first degree aggressiveness. There is also a lack of species with second degree aggressiveness, because of which the major portion of the area within the polar deserts is occupied by strongly disrupted, open aggregations on non-plakor sites, as mentioned in numerous instances above (p. 62, etc.). On Zemlya Aleksandra, where the vegetation has been studied most thoroughly, it is possible to distinguish a number of rather widely distributed and hemistenotopic species, but these are met with only as individual specimens over the greater part of this territory and belong therefore to the fourth or fifth degree of aggressiveness (i.e., species which are hardly, or not at all aggressive). Some species of cryptogams, which are widely distributed or hemi-stenotopic, are found everywhere, although in general only in small numbers (no species occurs in abundance) and may be referred to as being of the third degree of aggressiveness (i.e., they are species of medium aggressiveness). These include *Stereocaulon rivulorum*, *Ochrolechia frigida*, *Pertusaria glomerata*, *Cephaloziella arctica*, *Ditrichum flexicaule* and *Polytrichum alpinum*. Species growing in great abundance in various ecotopes within the limits of narrow ecological niches (e.g., *Umbilicaria proboscidea* and *Cetraria cucullata*) can at most be referred to as being of the fourth degree of aggressiveness (species which are not very aggressive and which are stenotopic in rare ecotopes). *Phippsia algida* appears to be the most aggressive of the flowering plants on Zemlya Aleksandra. The concept of the third degree of aggressiveness (i.e., species of medium aggressiveness) may be applied only to this flowering plant. *Papaver polare* can be referred to as being of the fourth degree of aggressiveness (hemi-stenotopic plants, few in number), as can *Saxifraga cernua*, *S. hyperborea*, *Draba oblongata* and *D. pauciflora*, which are widely distributed species growing as individual specimens. The rest of the flowering plants are non-aggressive.

Thus, the arctic polar deserts are also well distinguished from the arctic tundra in respect of the degree of aggressiveness of the species.

The plant aggregations which, because of their specific composition and structure, are classed as being of the arctic polar desert type of vegetation, form also the top-most belt on the mountains within the subzone of the arctic tundra. This vegetation, which has developed on 'plakor analogues', is of diagnostic significance for the recognition of altitudinal vegetation belts in the mountains found, according to the description furnished by Sochava (1956), especially on 'saddles, on gentle slopes and terraced surfaces, etc. ... on substrates formed mainly by fine soils and where, during winter, the vegetation is covered by snow, although not very thick but compact.' The transition into belts of the alpine polar deserts occurs at different altitudes, depending on the latitudinal position and the specificity of the climate. In particular, the characteristic of adiabatic winds is of major significance: if they take the form of a 'bora' (i.e., arctic north-easterlies, sometimes accompanied by blizzard conditions) such as occurs on the South Island of Novaya Zemlya, their cooling effect results in a lowering of the limit of the alpine polar-desert belt; in the case of foehn winds, which can heat up the mountain slopes (e.g., on Ostrov Vrangelya), it will raise the limit. The presence of fog or cold, wet winds blowing in from the sea, is also of importance. The effects of all these factors sometimes give rise to inversions of the atmospheric strata such as have been described as occurring on Ostrov Vrangelya (by Svatkov, 1970), on Peary Land (by Holmen, 1957) and on Axel Heiberg Island (by Beschel, 1963).

Brief accounts of the vegetation in the alpine polar deserts within the territory of the USSR have been published for Novaya Zemlya (by Aleksandrova, 1956), for Ostrov Kotel'niy (by Gorodkov, 1936; Korotkevich, 1958), for Ostrov Bol'shoy Lyakhovskiy (by Aleksandrova, 1963) and for Ostrov Vrangelya (by Gorodkov, 1958*b*). Alpine polar deserts in the Arctic outside the USSR have also been briefly described, for Devon Island (by Svoboda, 1977; Bliss, 1977), from Axel Heiberg Island (by Beschel, 1963), for Greenland (by Holmen, 1957; Schwarzenbach, 1961, etc.) and for Spitsbergen (by Hadač, 1946; Eurola, 1968).

Summary

The study of the vegetation of the arctic polar deserts is of major interest not only from a phytogeographical point of view, but also from a general biological point of view because it provides an opportunity for examining the pathways of adaptation of living organisms under extreme conditions of existence with respect to a pronounced deficiency of heat. Thanks to the results of new investigations within the areas of Mys

Chelyuskin, the Severnaya Zemlya and Zemlya Frantsa-Iosifa, and also due to the availability of more detailed material collected by myself on Zemlya Aleksandra, opportunities have arisen recently for making comparative analyses of the vegetation and the floras in different areas within the Soviet arctic polar deserts. These have enabled us to delimit certain areas more exactly, to analyse the structural characteristics of the plant communities in the polar deserts more carefully, to clarify the methods for the classification of these plant communities and to characterize the peculiarities of the biology and morphology of the plants. However, the vegetation of the polar deserts so far studied is still not adequately known, and future investigations ought to furnish much new information concerning the existence of life in the extreme northern areas at high latitudes in the Arctic.

Bibliography

Abramov, I. I., Abramova, A. L., Afonina, O. M. and Blagodatskikh, L. S. (1975). Arkticheskaya flora mkhov SSSR i yeye osobennosti. [The arctic bryophyte flora of the USSR and its peculiarities.] *Novosti sistematiki nizshikh rasteniy* [Nov. Syst. Pl. non Vascul.], vol. 12, 273–83.

Abramova, A. L., Savich-Lyubitskaya, L. I. and Smirnova, Z. N. (1961). *Opredelitel' listostebel'nikh mkhov Arktiki SSSR.* [Key to the bryophytes of arctic USSR.] Moscow, Leningrad.

Addison, P. A. and Bliss, L. C. (1980). Summer climate, microclimate, and energy budget of a polar semidesert on King Christian Island, N.W.T., Canada. *Arct. and Antarct. Res.* **12** (2), 161–70.

Afonina, O. M. (1974). Brioflora Chukotskogo polyostrova. [The bryoflora of the Chukot Peninsula.] Autoref., diss. 21 pp. Leningrad.

Aleksandrova, V. D. (1956). Rastitel'nost' Yuzhnogo ostrova Novoy Zemli mezhdu 70°56′ and 72°12′ s. sh. [The vegetation of the South Island of Novaya Zemlya between Lat. 70°56′ and 72°12′ N.] *Rastitel'nost' Kraynego Severa SSSR i yeye osvoyeniye* [The vegetation of the far north of the USSR and its utilization.], vol. 2, 187–306. Moscow, Leningrad.

Aleksandrova, V. D. (1960). Fenologiya rasteniy i sezonnyye aspekty v podzone arkticheskikh tundr. [The phenology of plants and seasonal aspects within the subzone of the arctic tundra.] *Trudy Fenologicheskogo soveshchaniya* [Papers of the phenological conference], pp. 209–18. Leningrad.

Aleksandrova, V. D. (1961). Vliyaniye snezhnogo pokrova na rastitel'nost' v arkticheskoy tundre Yakutii. [Effects of snow cover on vegetation in the arctic tundra of Yakutia.] *Materialy po rastitel'nosti Yakutii* [Materials concerning the vegetation of Yakutia], pp. 190-221. Leningrad.

Aleksandrova, V. D. (1962). O podzemnoy strukture nekotorykh rastitel'nykh soobshchestv arkticheskoy tundry o. B. Lyakhovskogo. [On the subterranean structure of some plant associations in the arctic tundra of Ostrov Bol'shogo Lyakhovskogo.] *Problemy botaniki* [Botanical problems], vol. 6, 148–60. Moscow, Leningrad.

Aleksandrova, V. D. (1963). Ocherk flory i rastitel'nost' o. Bol'shogo Lyakhovskogo. [Essay on the flora and vegetation of Ostrov Bol'shogo Lyakhovskogo.] *Trudy Arkt. i Antarkt. Inst.* [Papers of the Arctic and Antarctic Institute] **224**, 6–36.

Aleksandrova, V. D. (1969). Nadzemnaya i podzemnaya massa rasteniy polyarnoy pustyni ostrova Zemlya Aleksandry (Zemlya Frantsa Iosifa). [The above-

and below-ground mass of plants in the polar deserts of the island of Zemlya Aleksandra (Zemlya Frantsa-Iosifa).] *Problemy Botaniki* (Botanical Problems), vol. 11, 47–60. Leningrad.

Aleksandrova, V. D. (1971). Printsipy zonal'nogo deleniya rastitel'nosti Arktiki. [Principles for the zonal division into areas of the vegetation of the Arctic.] *Bot. Zhurn.* **56** (1), 3–21.

Aleksandrova, V. D. (1977*a*). Geobotanicheskoye rayonirovannie Arktiki i Antarktiki. [Geobotanical division into areas of the Arctic and the Antarctic.] *Komarovskiye chteniya* [Komarov Lecture Series], no. 29, 188 pp. Leningrad.

Aleksandrova, V. D. (1977*b*). Struktura rastitel'nykh gruppirovok polyarnoy pustyni o. Zemlya Aleksandry (Zemlya Frantsa-Iosifa). [The structure of the plant aggregations of the polar deserts on the island of Zemlya Aleksandra (Zemlya Frantsa-Iosifa).] *Problemy ekologii, geobotaniki, botanicheskoy geografii i floristiki* [Problems of ecology, geobotany, botanical geography and floristics], pp. 26–36. Leningrad.

Aleksandrova, V. D. (1979). Proyekt klassifikatsii rastitel'nosti Arktiki. [Project for the classification of arctic vegetation.] *Bot. Zhurn.* **64** (12), 1715–30.

Aleksandrova, V. D. (1980). *The Arctic and the Antarctic: their division into geobotanical areas.* 247 pp. Cambridge, England.

Aleksandrova, V. D. (1981). Otkrytyye rastitel'nye gruppirovki polyarnoy pustyni o. Zemlya Aleksandry (Zemlya Frantsa-Iosifa) i ikh klassifikatsiya. [Open plant aggregations on the island of Zemlya Aleksandra (Zemlya Frantsa-Iosifa) and their classification.] *Bot. Zhurn.* **66** (5), 636–49.

Andreyev, M. P. (1981). Lishayniki rodov *Cladina* i *Cladonia* Anyuyskogo nagor'ya. [The lichen genera *Cladina* and *Cladonia* in the Anyuy foothills.] *Bot. Zhurn.* **66** (1), 31–41.

Arkticheskaya Flora SSSR [The arctic flora of the USSR]. 1960–80: no. 1, 102 pp., 1960; no. 2, 273 pp., 1964; no. 3, 175 pp., 1966; no. 4, 96 pp., 1963; no. 5, 208 pp., 1966; no. 6, 247 pp., 1971; no. 7, 180 pp., 1975; and no. 8, 332 pp., 1980. Leningrad.

Bell, K. L. and Bliss, L. C. (1978). Root growth in a polar semidesert environment. *Can. J. Bot.* **56** (20), 2470–90.

Berg, L. S. (1928). Zona tundr. Opyt landshaftnoy characteristiki. [The tundra zone. Experiments in landscape characterization.] *Izv. Leningr. univ.* [Publications from the University of Leningrad] **1**, 191–233.

Berg, L. S. (1936). *Fisiko-geograficheskiye (landshaftnyye) zony SSSR.* [The physical-geographical (landscape) zones of the USSR], 2nd edn, part 1. 427 pp. Leningrad.

Beschel, R. E. (1963), Geobotanical studies on Axel Heiberg Island in 1962. *Axel Heiberg Island. Preliminary report 1961–1962,* ed. T. Müll, pp. 199–215. Montreal.

Beschel, R. (1969). Floristicheskiye sootnosheniya na ostrovakh Neoarktiki. [Floristic correlation between the Neo-arctic islands.] *Bot. Zhurn.* **54** (6), 872–91.

Birulya, A. (1907). Ocherki iz shizni ptits polyarnogo poberezh'ya Sibiri. [Essays on the bird life of the Siberian polar coast.] *Zap. Akad. Nauk.*, ser. 8 **18** (2), 1–53.

Blagodatskikh, L. S. (1973). Listostebel'nye mkhi rayona Taymyrskogo statsionary (Zapadnyy Taymyr). [The bryophytes of the area around the Taymyr experimental station (Western Taymyr).] *Biogeotsenozy Taymyrskoy tundry i ikh produktivnost'* [Biogeocoenoses of the Taymyr tundra and their productivity], no. 2, 107–19. Leningrad.

Blagodatskikh, L. S., Zhukova, A. L. and Matveyeva, N. V. (1979). Listotel'nyye i

pechenochnyye mkhi mysa Chelyuskin. [Bryophytes and hepatics on Mys
Chelyuskin.] *Arkticheskiyye tundry i polyarnyye pustyni Taymyra* [The arctic
tundra and polar deserts of Taymyr], pp. 54–60. Leningrad.

Bliss, L. C. (1975). Tundra grasslands, herblands, and shrublands and the role of
herbivores. *Geosci. and Man* **10**, 51–79.

Bliss, L. C. (1977). Introduction. *Truelove lowland, Devon Island, Canada: a high
Arctic ecosystem.* pp. 1–11. Edmonton, Alberta, Canada.

Böcher, T. W. (1977). *Cerastium alpinum* and *C. arcticum*, a mature polyploid
complex. *Bot. Notiser* **130**, 303–9.

Böcher, T. W., Holmen, K. and Jakobsen, K. (1968). *The flora of Greenland.* 312
pp. Copenhagen.

Byalynitskiy-Birulya, A. (1902). Otchet o botanicheskikh issledovanniyakh za
letniy sezon 1901 g. [Account of botanical investigations during the summer
season of 1901.] *Izv. Akad. Nauk.* **16** (5), 226–7.

Chernov, Yu. I., Striganova, B. P., Anan'yeva, S. I. and Kuz'min, L. L. (1979).
Zhivotniy mir polyarnoy pustyni mysa Chelyuskin. [Animal life of the polar
deserts on Mys Chelyuskin.] *Arkticheskiye tundry i polyarnyye pustyni Taymyra*
[The arctic tundra and polar deserts of Taymyr], pp. 35–49. Leningrad.

Chugunova, M. V. (1979). Nekotoryye khimicheskiye svoystva pochv mysa
Chelyuskin. [Some chemical properties of the soils of Mys Chelyuskin.]
Arkticheskiye tundry i polyarnyye pustyni Taymyra [The arctic tundra and polar
deserts of Taymyr], pp. 74–7. Leningrad.

Dibner, V. D. (1961). Zarosli ivy mokhnatoy za 75-y parallel'yu. [A *Salix lanata*
thicket above the 75th parallel.] *Izd. BGO,* **93**, (4) 334–336.

Dibner, V. D. (1965). Istoriya formirovaniya pozdnepleystotsenivikh i golotseno-
vikh otlozheniy Zemlya Frantsa-Iosifa. [The history of the formation of the late
pleistocene and holocene deposits on Zemlya Frantsa-Iosifa.] *Tr. Nauch.-issl.
instituta Geologii Arktiki* [Papers of the Institute for Arctic Geology] **143**,
300–18.

Dibner, V. D. (1970). *Geomorfologiya.* [Geomorphology.] *Sovetskaya Arktika*
[The Soviet Arctic], pp. 59–93. Moscow.

Dibner, V. D., Gakkel', Ya. Ya., Litvin, V. M., Martynov, V. T. and Shugayeva,
N. D. (1965). Geomorfologicheskaya karta Severnogo Ledovitogo okeana.
[Geomorphological map of the Arctic Ocean.] *Antropogenovyy period v Arktike
i Subarktike* [The anthropogenic period in the Arctic and the Subarctic], pp.
341–5. Moscow.

Esipov, V. K. (1933). Arkhipelag Zemlya Frantsa-Iosifa. Fiziko-geograficheskiy
ocherk. [The archipelago of Zemlya Frantsa-Iosifa. A physical-geographical
essay.] *Ostrova Sovetskoy Arktiki* [The Soviet Arctic Islands], pp. 116–35.
Arkhangelsk.

Eurola, S. (1968). Ueber die Fieldheidevegetation in den Gebieten von Isfjorden
and Hornsund in Westspitzbergen. [On the vegetation of the montane heaths in
the areas of Isfjord and Hornsund in Western Spitsbergen.] *Aquilo,* Ser. Bot. 7,
1–56.

Eurola, S. (1971). Die Vegetation einer Sturzhalde (Sveagruva, Spitsbergen,
77°53′ no. Br.). [The vegetation of a scree (at the Svea mine, Spitsbergen, N. lat.
77°53′).] *Aquilo,* Ser. Bot. **10**, 8–28.

Fischer, H. (1896). Some remarks on the flora of Frantz Josef Archipelago. *Geogr.
J.* **8**, 560–3.

Fiziko-geograficheskiy Atlas Mira (1964). [The physical-geographical atlas of the
world.) 298 pp. Moscow.

Flovik, K. (1940). Chromosome numbers and polyploidy within the flora of
Spitzbergen. *Hereditas* **26**, 430–40.

212   Bibliography

Gorodkov, B. N. (1935). *Rastitel'nost' tundrovoy zony SSSR*. [Vegetation of the tundra zone of the USSR.] 142 pp. Leningrad.

Gorodkov, B. N. (1956). Rastitel'nost' i pochvy o. Kotel'nogo (Novosibirskiy Arkipelag). [Vegetation and soils on Ostrov Kotel'niy (in the Novosibirskiy Archipelago).] *Rastitel'nost' Kraynego Severa SSSR i yeye osvoyeniye* [The vegetation of the Soviet Far North and its utilization], no. 2, 7–132. Moscow, Leningrad.

Gorodkov, B. N. (1958*a*). Analiz rastitel'nosti zony arkticheskikh pustyn' na primere ostrova Vrangelya. [Analysis of the vegetation of the arctic deserts, using Ostrova Vrangelya as an example.] *Rastitel'nost' Kraynego Severa SSSR i yeye osvoyeniye* [The vegetation of the Soviet Far North and its utilization], no. 3, 59–94. Moscow, Leningrad.

Gorodkov, B. N. (1958*b*). Pochvenno-rastitel'nostyy pokrov ostrova Vrangelya. [The soil and vegetation cover on Ostrov Vrangelya.] *Rastitel'nost' Kraynego Severa SSSR i yeye osvoeyniye* [The vegetation of the Soviet Far North and its utilization], no. 3, 5–58. Moscow, Leningrad.

Govorenkov, B.F.O. (1981). O pochvoobrazovanii v usloviyakh arkticheskikh pustyni o-va Oktyabr'skoy Revolyutsii. [On soil formation under arctic desert conditions on Ostrov Oktaybr'skoy Revolyutsii.] *Tr. Arkt. i Antarkt. Inst.* 367, 132–41.

Govorukha, L.S. (1960). K mikroflore Zemli Frantsa-Iosifa. [On the microflora of Zemlya Frantsa-Iosifa.] *Problemy Arktiki i Antarktiki* [Problems of the Arctic and the Antarctic], no. 3, 119–21. Leningrad.

Govorukha, L.S. (1968). Landshaftno-geograficheskaya kearakteristika Zemli Frantsa-Iosifa. [A landscape-geographical characterization of Zemlya Frantsa-Iosifa.] *Tr. Arkt. i Antarktiki* **285**, 86–117.

Govorukha, L.S. (1970*a*). Zemlya Frantsa-Iosifa. *Sovetskaya Arktika* [The Soviet Arctic], pp. 328–59. Moscow.

Govorukha, L.S. (1970*b*). Ostrov Viktoriya. *Sovetskaya Arktika* [The Soviet Arctic], pp. 359–63. Moscow.

Govorukha, L.S. (1976). Severnaya Zemlya. *Bol'shaya Sovetskaya Entsiklopediya* **23**, p. 124.

Grigor'yev, A.A. (1946). *Subarctica*. [The Subarctic.] 171 pp. Moscow, Leningrad.

Hadač, E. (1946). The plant communities of Sassen Quarter, Vestspitsbergen. *Stud. Bot. Cechoslovaca* **7** (2–3), 127–64.

Hanssen, O. and Lid, J. (1932). Flowering plants of Franz Josef Land. *Skr. Svalbard og Ishavet*, no. 39, 1–42.

Hedberg, O. (1967). Cytotaxonomic studies on *Anthoxanthum odoratum* L. *s. lat.* II *Symb. Bot. Upsal.* **8** (5), 1–88.

Holmen, K. (1957). The vascular plants of Peary Land, North Greenland. *Medd. Grønland* **124** (9), 1–154.

Hultén, E. (1958). *The amphiatlantic plants and their phytogeographical connections*. 340 pp. Stockholm.

Hultén, E. (1968). *Flora of Alaska and neighboring territories*. 1008 pp. Stanford, CA.

Johnson, A.W. and Packer, J.G. (1968). Chromosome numbers in the flora of Ogotoruk Creek, S.W. Alaska. *Bot. Notiser* **121**, 403–56.

Karavayeva, N.A. and Targul'yan, V.O. (1978). Avtonomnoye pochvoobrazovaniye na severe Yevrazii i Ameriki. [Autonomous soil formation in northern Eurasia and America.] *Problemy pochvovedeniya* [Problems of soil science], pp. 174–8. Moscow.

Kartushin, V.M. (1963*a*). Oledeneniye o. Bennetta. [Glaciation on Ostrov Bennetta.] *Tr. Arkt. i Antarkt. Inst.* **224**, 166–176.

Kartushin, V.M. (1963*b*). O rastitel'nosti o. Bennetta. [On the vegetation of Ostrov Bennetta.] *Tr. Arkt. i Antarkt. Inst.* **224**, 177–9.

Khodachek, E.A. (1970). Semennaya productivnost' i urozhay semyan rasteniy v tundrakh Zapadnogo Taymyra. [Seed production and seed yield in the tundra of western Taymyr.] *Bot. Zhurn.* **55** (7), 995–1010.

Khodachek, E.A. (1980). O rastitel'nosti polyostrova Zhilogo (o. Oktyabr'skoy Revolyutsii, Arkhipelag Severnaya Zemlya). [On the vegetation of Zhiloy Peninsula (Ostrov Oktyabr'skoy Revolyutsii, the archipelago of Severnaya Zemlya).] Manuscript, Bot. Inst. Akad. Nauki SSSR.

Kjellman, F.R. (1883*a*). Die Phanerogamenflora der Sibirischen Nordküste. [The phanerogam flora of the Siberian northern coast.] *Die wissenschaftlichen Ergebnisse der Vega-Expedition* [The scientific results of the "Vega" expedition], vol. 1, 94–139. Leipzig.

Kjellman, F.R. (1883*b*). Üeber den Pflanzenwuchs an der Nordküste Sibiriens. [On the vegetation of the Siberian northern coast.] *Die wissenschaftlichen Ergebnisse der Vega-Expedition* [The scientific results of the "Vega" expedition], vol. 1, 80-93. Leipzig.

Korotkevich, Y.S. (1958). Rastitel'nost' Severnoy Zemli. [The vegetation of Severnaya Zemlya.] *Bot. Zhurn.* **43** (5), 644–63.

Korotkevich, Y.S. (1967). Polyarnyye pustyni. [Polar deserts.] *Byull. Sov. Antarkt. ekpeditsii* [Bulletin of the Soviet Antarctic Expedition], no. 65, 5–29.

Korotkevich, Y.S. (1972). *Polyarnyye pustyni.* [Polar deserts.] 420 pp. Leningrad.

Ladyzhenskaya, K.I. and Zhukova, A. 01971). O morfologicheskikh osoben-nostyakh pechenochnykh mkhov v usloviyakh visokoshirotnoy Arktiki. [On the morphological peculiarities of the liverworts under the conditions of the high-latitude Arctic.] *Ekologiya*, no. 3, 26-30.

Ladyzhenskaya, K.I. and Zhukova, A. (1972). Pechenochnye mkhi (Hepaticae) ostrova Zemlya Aleksandry. [The liverworts (Hepaticae) of the island of Zemlya Aleksandra.] *Bot. Zhurn* **57** (3), 348–53.

Lavrenko, Y.M. (1947). Printsipy i yedinitsy geobotanicheskogo rayonirovaniya. [Principles and units of geobotanical division into areas.] *Geobotanicheskoye rayonirovaniye SSSR* [Geobotanical division into areas of the USSR], pp. 9-13. Moscow, Leningrad.

Lavrenko, Y.M. (1959). Osnovnyye zakonomernosti rastitel'nykh soobshchestv i puti ikh izucheniya. [Regularity of plant associations and methods for studying them.] *Polevaya geobotanika* [Field manual of geobotany], vol. 1, 13–75. Moscow, Leningrad.

Lavrenko, Y.M. (1968). Ob ocherednykh zadachakh izucheniya geografii rastitel'-nogo pokrova v svyazi s botaniko-geograficheskim rayonirovaniyem SSSR. [On the present need for a study of the geography of the plant cover in connection with the geographical division of the USSR into areas.] *Osnovyye problemy sovremennoy geobotaniki.* [Basic problems of modern geobotany], pp. 45-68. Leningrad.

Leonov, L.I. (1944). Geomorfologicheskiy ocherk ostrova Genrietty. [Geomor-phological essay on Ostrov Genrietty.] *Problemy Arktiki*, no. 1, 133–148.

Leonov, L.I. (1953). *V vysokikh shirotakh (zapiski naturalista).* [On high latitudes (notes by a naturalist).] 123 pp. Moscow.

Leskov, A.I. (1947). Arkticheskaya tundrovaya oblast'. [The region of arctic tundras.] *Geobotanicheskoye rayonirovanie SSSR* [The geobotanical division of the USSR into areas], pp. 14-17. Moscow, Leningrad.

Levkovskiy, V.N., Tikhmenev, Y.A. and Levkovskiy, Y.P. (1981). Kleystogamiya arkticheskikh zlakov. [Cleistogamy of arctic grasses.] *Bot. Zhurn.* **66** (1), 116–20.

Lindsay, D.C. (1975). The macrolichens of South Georgia. *Brit. Antarct. Sci. Reports*, no. 89, 1–91.

Lindsay, D.C. (1977). Lichens of cold deserts. *Lichen ecology*, pp. 183–209. London.

Löve, 'A. (1977). Islenzk ferdaflóra. [Icelandic excursion flora]. 429 pp. Reykjavík, Iceland.

Löve, 'A. and Löve, D. (1975). *Cytotaxonomical atlas of the Arctic flora*. 598 pp. Lichtenstein.

Lynge, B. (1928). Lichens from Novaya Zemlya. *Rep. Sci. Res. Norweg. Expedition to Novaya Zemlya 1921*, no. 43, 1–270. Kristiania, Norway.

Lynge, B. (1929). Vascular plants and lichens. *The Norwegian North Polar expedition with the "Maud", 1918–1925, Sci. res.*, vol. 5 (1), 3–15. Bergen. Norway.

Lynge, B. (1931). Lichens collected on the Norwegian scientific expedition to Franz Josef Land. *Skr. Svalbard og Ishavet*, no. 38, 3–18.

Matveyeva, N.V. (1979). Struktura rastitel'nogo pokrova polyarnykh pustyn' polyostrova Taymyr (mys Chelyuskin). [The structure of the plant cover in the polar deserts of the Taymyr peninsula (Mys Chelyuskin).] *Arkticheskiye tundry i polyarnyye pustyni Taymyra* [The arctic tundras and polar deserts of Taymyr], pp. 5–27. Leningrad.

Matveyeva, N.V. and Chernov, Yu. I. (1976). Polyarnyye pustyni polyostrova Taymyr. [The polar deserts of Taymyr peninsula.] *Bot. Zhurn.* 61 (3), 297–312.

Matveyeva, N.V., Polozova, T.G., Blagodatskikh, L.S. and Dorogostaiskaya, E.V. (1973) Kratkiy ocherk rastitel'nosti Taymyrskogo biogeotsenologiches-kogo statsionara. [Short list of the vegetation at the Taymyr biological station.] *Biogeotsenozy Taymyrskoy tundry i ikh produktivnost'* [Biogeocoenoses of the Taymyr tundras and their productivity], vol. 2, pp. 7–29.

Middendorf, A.F. (1860–7). Puteshestviye na sever i vostok Sibiri. [An expedition to northern and eastern Sibera.] *SPb.*, part 1. 758 pp.

Mikhayilov, I.S. (1960). Nekotorye osobennosti dernovykh arkticheskikh pochv o. Bol'shevik. [Some peculiarities of ancient arctic soils on Ostrov Bol'shevik.] *Pochvovedeniye* [Soil science], no. 6, 89–92.

Mikhaylov, I.S. (1962). Pochvi polyarnykh pustin' i rol' B.N. Gorodkova v ikh izuchenii. [The soils of the polar deserts and the role of B.N. Gorodkov in their investigation.] *Izv. VGO* (Journal of the Geographical Division) 94 (6), 520–3.

Mikhaylov, I.S. (1970): Pochvy. [Soils.] *Sovetskaya Arktika* [The Soviet Arctic], pp. 236–49. Moscow.

Mikhaylov, I.S. and Govorukha, L.S. (1962). Pochvy Zemli Frantsa-Iosifa. [The soils of Zemlya Frantsa-Iosifa.] *Vestn. Mosk. Univ.* (Journal of the University of Moscow), Ser. 5 (Geography), no. 6, 42–8.

Miroshnikov, L.D. (1973). O svyazi geograficheskogo pasprostraneniya rasteniy na poluostrove Chelyuskin s geologicheskimi faktorami. [The correlation between geographical distribution of the plants on Mys Chelyuskin and geological factors.] *Izv. VGO* 105 (1), 40–2.

Norin, B.N. (1970). O funktsional'noy strukture rastitel'nykh gruppirovok leso-tundry. [On the functional structure of plant aggregations of the forest tundra ecotone.] *Bot. Zhurn.* 55 (2), 170–83.

Norin, B.N. (1979). *Struktura rastitel'nykh soobshchestv vostochnoyevropeyskoy lesotundr.* [The structure of the plant associations of the east-European forest tundra ecotone.] 200 pp. Leningrad.

Novichkova-Ivanova, L.N. (1963). Smeny sinuziy pochvenikh vodorosley Zemli Frantsa-Iosifa. [Changes in the synusia of soil algae on Zemlya Frantsa-Iosifa.] *Bot. Zhurn.* 48 (1), 42–53.

Novichkova-Ivanova, L.N. (1972). Soil and aerial algae of polar deserts and arctic tundra. *Tundra biome: Proceedings* IV *International meeting on the biological productivity of tundra*, ed. F.E. Wielgolaski and Th. Rosswall, pp. 261–5. Stockholm.

Oksner, A.N. (1944). O proiskhozhdenii areala bipolyarnkh lishaynikov. [On the origin of the areas of distribution of the bipolar lichens.] *Bot. Zhurn.* **29** (6), 243–56.

Palibin, I.V. (1903). Botanicheskiye razul'taty plavaniya ledokola 'Ermak' v Severnom Ledovitom Okeane letom 1901 g. [Botanical results of the voyage of the icebreaker 'Yermak' through the North Polar Ocean during the summer of 1901.] *Izv. S.-Peterbugr. Bot. Sad.* [Publ. of the St Petersburg Botanical Garden] **3** (5), 135–67.

Parinkina, O.M. (1979). Mikroflora pochv polyarnykh pustyn' mysa Chelyuskin. [The soil microflora of the polar desert on Mys Chelyuskin.] Arkticheskiye tundry i polyarnyye pustyni Taymyra [The Arctic tundra and polar deserts of Taymyr], pp. 28-34. Leningrad.

Passarge, S. (1920). *Die Grundlagen der Landschaftskunde.* [The basics of landscape investigation.] 588 pp. Hamburg.

Perfil'yev, I.A. (1928). *Materialy k flore ostrovov Novoy Zemli i Kolguyeva.* [Material for a flora of the islands of Novaya Zemlya and Kolguyev.] 74 pp. Arkhanglesk.

Petrovskiy, V.V. and Zhukova, P.G. (1981). [Chromosome numbers and taxonomy of some plant species on Ostrov Vrangelya.] *Bot. Zhurn.* **66**; 380–7.

Piyn, T. Kh. (1979). Nanopochvennyye lishayniki mysa Chelyuskin. [Soil nano-lichens of Mys Chelyuskin.] *Arkticheskiye tundry i polyarnyye pustyni Taymyra* [The arctic tundras and polar deserts of Taymyr], pp. 61-73. Leningrad.

Piyn, T. Kh. and Trass, Kh. Kh. (1971). Nanopochvennyye lishayniki okrestnostey Tarei. [Soil nano-lichens of the Tareya area.] *Biogeotsenozy Taymyrskoy tundry i ikh produktivnost'* [The biogeocoenoses of the Taymyr tundra and their productivity], pp. 151–60. Leningrad.

Polozova, T.G. (1978). Zhizhnennye formy sosudistikh rasteniy Taymyrskogo stationara. [Life-forms of the vascular plants at the Taymyr Experimental Station.] *Struktura i funktsii biogeotsenozov Taymyrskogo tundry* [The structure and function of the biogeocoenoses of the Taymyr tundra], pp. 114–43. Leningrad.

Polozova, T.G. and Tikhomirov, B.A. (1971). Sosudistyye rasteniya rayona Taymyrskogo statsionara (pravoberezh'ye Pyasiny bliz yst-ya Tarei, zapadniy Taymyr). [The vascular plants of the area around the Taymyr Experimental Station (on the right-hand bank of Pyasina river near the mouth of river Tareya).] *Struktura i funktsii biogeotsenozov Taymyrskogo tundry* [The structure and function of the biogeocoenoses in the Taymyr tundral], pp. 161–84. Leningrad.

Porsild, A.E. (1955). The vascular plants of the Western Canadian Arctic Archipelago. *Nat. Mus. Can. Bull.* no. 135. 1–226.

Porsild, A.E. (1957). Illustrated flora of the Canadian Arctic Archipelago. *Nat. Mus. Can. Bull.*, no. 146, 1–209.

Prik, Z.M. (1970). Klimat. [The climate.] In: *Sovetskaya Arktika* [The Soviet Arctic], pp. 108–47. Moscow.

Rachkovskaya, E.I. (1977). Kraynearidnyye tipy pustyn' v Zaaltayskoy Gobi. [Extremely arid types of deserts in the Cis-Altaian Gobi desert.] *Problemy ekologii, geobotaniki, botanicheskoy geografii i floristiki* [Problems of ecology, geobotany, botanical geography and floristics], pp. 99–108. Leningrad.

Regel, C. (1932). Die Fleckentundra von Nowaja Semlja. [The 'spotted' tundra of

Novaya Zemlya.] *Beitr. Biol. Pflanz.*, vol. *20*, 7–12.

Safronova, I.N. (1976). K flore i rastitel'nosti o. Oktyabr'skoy Revolyutsii (arkhipelag Severnaya Zemlya). [On the flora and vegetation of Ostrov Oktyabr'skoy Revolyutsii (in the archipelago of Severnaya Zemlya).] *Biologicheskiye problemy Severa* [Biological problems of the North], VIIth Symposium (Botany), pp. 191–3. Petrozavodsk. (Abstracts of lectures.)

Safronova, I.N. (1979). Sosudistyye rasteniya mysa Chelyuskin. [The vascular plants of Mys Chelyuskin.] *Arkticheskiye tundry i polyarnyye pustyni Taymyra* [The arctic tundra and polar deserts of Taymyr], pp. 50–53. Leningrad.

Safronova, I.N. (1981*a*). O flore i rastitel'nosti o. Gukera i o. Meybel (arkhipelag Zemlya Frantsa-Iosifa). [On the flora and vegetation of Hooker and Mabel islands (the archipelago of Zemlya Frantsa-Iosifa).] *Biologicheskiye Problemy Severa* [Biological problems of the North], VIIIth Symposium. 44 pp. Syktyvkar. (Abstract of lecture.)

Safronova, I.N. (1981*b*). Flora ostrova Oktyabr'skoy Revolyutsii. [The flora of Ostrov Oktyabr'skoy Revolyutsii.] *Tr. Arkt. i Antarkt. Inst.* **367**, 142–50.

Safronova, I.N. (1983). Materialy k flore o. Meybel i o. Gukera. [Material for a flora of Mabel and Hooker Islands.] *Bot. Zhurn.* **68** (4) 513–19.

Sambuk, F.V. (1937). Kratkiy ocherk rastitel'nosti Taymyra. [A short essay on the vegetation of Taymyr.] *Problemy Arktiki* [Problems of the Arctic], no. 1, 127–53.

Samoylovich, R.L. (1937). Ocherk geomorfologii Novoy Zemli. [Essay on the geomorphology of Novaya Zemlya.] *Novozemel'skaya ekskursiya* [The expedition to Novaya Zemlya], part 1, 76–101. Leningrad.

Savich, L.I. (1936). Mkhi arkhipelaga Fratsa-Iosifa, Severnoy Zemli i o. Vize, sobrannyye V.P. Savichem vo vremya polyarnoy ekspeditsii 1930 g. na ledokole 'G. Sedov'. [The mosses of Zemlya Frantsa-Iosifa, Severnaya Zemlya and Ostrov Vize, collected by V.P. Savich during the polar expedition on board the icebreaker 'G. Sedov' in 1930.] *Tr. BIN AH SSSR* [Papers from the Botanical Institute of the Academy of the Sciences of the USSR], ser. II (Cryptogames), no. 3, 505–78.

Schwarzenbach, F.H. (1961). Botanische Beobachtungen in der Nunatakerzone Ostgrönlands zwischen 74° and 75° n. Br. [Botanical observations in the nunatak zone of East Greenland between Lat 74° and 75° N.] *Medd. Grönland* **163** (5), 1–172.

Semenov, I.V. (1966*a*). Vnutrilandshaftnoye rayonirovaniye Severnoy Zemli. [Division of the interior landscapes of Severnaya Zemlya into areas.] *Izv. VGO,* **98** (6), 477–85.

Semenov, I.V. (1966*b*). Fiziko-geograficheskoye rayonirovanie Severnoy Zemli. [Physical-geographical division of Severnaya Zemlya into areas.] *Izv. VGO* **98** (1), 3–9.

Semenov, I.V. (1968). O zakonomernostyakh differentsatsii prirodnikh usloviy ostrovov Sovetskoy Arktiki. [On regularity in the differentiation of natural conditions on the islands of the Soviet Arctic.] *Tr. Arkt. i Antarkt. Inst.* **285**, 74–85.

Semenov, I.V. (1970). Severnaya Zemlya. *Sovetskaya Arktika* [The Soviet Arctic], pp. 391–421. Moscow.

Semenov, I.V. (1981*a*). Morfologia lednikov Severnoy Zemli. [The morphology of the glaciers on Severnaya Zemlya.] *Tr. Arkt. i Antarkt. Inst.* **367**, 9–20.

Semenov, I.V. (1981*b*). Prostranstvennyye zakonomernosti razvitiya oledeneniya Severnoy Zemli. [Spatial regularity of the distribution of glaciers on Severnaya Zemlya.] *Tr. Arkt. i Antarkt. Inst.* **367**, 21–30.

Shamurin, V.F. (1966a). Arkticheskiye pustyni i tundrovaya zona. [Arctic deserts and the tundra zone.] *Metody fenologicheskikh nablyudeniy pri botanicheskikh issledovaniyakh* [Methods of phenological observations for botanical research], pp. 24–33. Moscow, Leningrad.

Shamurin, V.F. (1966*b*). Sezonnyy ritm i ekologiya tsveteniya rasteniy tundrovikh soobshchestv na severe Yakutii. [Seasonal rhythm and ecology of the flowering plants in tundra associations of northern Yakutia.] *Prisposobleniye rasteniy Arktiki k usloviyam sredy* [The adapation of the plants in the Arctic to the conditions of their environment], pp. 5–125. Moscow, Leningrad.

Shennikov, A.P. (1964). *Vvedeniya v geobotaniki.* [Introduction to geobotany.] 447 pp. Leningrad.

Shul'ts, G.E. (1966). Voprosy metodiki i ogranizatsii fitofenologicheskikh nablyudeniy. [Problems of methodology and organization of phytophenological data.] In: Metodiki fenologicheskikh nablyudeniy pri botanicheskikh issledovaniyakh [Methods of phenological observations for botanical research], pp. 5–23. Moscow, Leningrad.

Sisko, R.K. (1970). Novosibirskiy arkhipelag. [The Novosiberian Islands.] In: *Sovetskaya Arktiki* [The Soviet Arctic], 422–52. Moscow.

Sochava, V.B. (1948). Geograficheskie svyazi rastitel'nogo pokrova na territorii SSSR. [The geographical correlations of the plant cover of the USSR.] *Uchen. zap. Leningr. ped. inst. A.I. Gertsena* [Scientific reports from the A.I. Herzen pedagogical institute, Leningrad] **73** (from the chair of physical geography), 3–51.

Sochava, V.B. (1956). Zakonomernosti geografii rastitel'novo pokrova gornykh tundr SSSR. [The regularities of the geography of the vegetation of alpine tundras in the USSR.] *Akademiky V.N. Sukachev k 75-letiyu so dnya rozhdeniya* [Festschrift for Academician V.N. Sukachev on the occasion of his 75th birthday], pp. 522–36. Moscow, Leningrad.

Sochava, V.B. and Gorodkov, B.N. (1956). Arkticheskie pustyni i tundry. [Arctic deserts and tundras.] *Rastitel'nyy pokrov SSSR* [The vegetation of the USSR], vol. 1, 61-139. Moscow, Leningrad.

Størmer, P. (1940). Bryophytes from Franz Josef Land and eastern Svalbard. *Norg. Ishav. Unders., Medd,* no. 47, 1–16.

Sukachev, V.N. (1975). Chto takoe fitotsenoz? [Why such a phytocoenosis?] In: *Izbrannie trudy, T. 3, Problemy fitotsenologii* [Collected works, vol. 3, Problems of phytocoenology], pp. 279–92. Leningrad.

Summerhayes, V.S. and Elton, C.S. (1928). Further contribution to the ecology of Spitsbergen. *J. Ecol.* **16** (2), 193–267.

Sushkina, N.N. (1960). Ob osobennostyakh mikroflory arkticheskikh pochv. [On the peculiarities of the microflora in arctic soils.] *Pochvovedenie* [Soil management], no. 4, 57–67.

Svatkov, N.M. (1970). Ostrov Vrangelya. In: *Sovetskaya Arktika* [The Soviet Arctic], pp. 433–72. Moscow.

Svoboda, J. (1977). Ecology and primary production of Raised Beach communities on Truelove Lowland. *Truelove lowland, Devon Island, Canada: A high Arctic ecosystem,* ed. L.C. Bliss, pp. 185–216. Edmonton, Canada.

Taymyro–Severozemel'skaya Oblast' (1970). [The Taymyr–Severnaya Zemlya Region] ed. R.K. Sisko, 373 pp. Leningrad.

Tedrow, J.C.F. (1970). Soil investigation in Inglefield Land, Greenland. *Medd. Grønland* **188** (3), 5-93.

Tedrow, J.C.F. (1972). Soil morphology as an indicator of climate changes in the Arctic areas. Acta univ. Ouluen. A (3), 61–74.

Tikhomirov, B.A. (1948a). K poznaniyu flory kraynikh polyarnykh predelov Evrazii. [To knowledge of the flora in the extreme polar areas of Eurasia.] *Byull.*

218   *Bibliography*

MOIP [Bull. of the Moscow Inst. of Pedagogy], Biol. Div. **53** (4), 91-102.
Tikhomirov, B.A. (1948*b*). K kharakteristike flory zapadnovo poberezh'ya Taymyra. [To the characterization of the flora on the western coast of Taymyr.] *Tr. Karelo-Fin. Un-ta* [Papers from the Karelo-Finnish University], **2**, 1–84.
Tkachenko, B.V. and Atlasov, I.P. (1970). Geologicheskaya struktura ostrovov i dna shel'fovykh morey. [The geological structure of the islands and the shelf bottom of the sea.] *Sovetskaya Arktika* [The Soviet Arctic], pp. 47–57. Moscow.
Tolmachev, A.I. (1931). Materialy dlya flory yevropeyskikh arkticheskikh ostrovov. [Material for a flora of the European arctic islands.] *Zhurn. Rus. bot. o-va* [J. Russian Bot. Assoc.] **16** (5–6), 459–72.
Tolmachev, A.I. (1936). Obzor flory Novoy Zemli. [Survey of the flora of Novaya Zemlya.] *Arctica*, no. 4, 143–78.
Tolmachev, A.I. (1959). K flore ostrova Bennetta. [On the flora of Ostrov Bennetta.] *Bot. Zhurn.* **44** (4), 543–5.
Tolmachev, A.I. (1971). *Cerastium* L. – Yaskolka. [*Cerastium* L. – Chickweed.] *Arkticheskaya Flora SSSR* [The arctic flora of the USSR], no. 6, 30–52.
Tolmachev, A.I. and Shukhtina, G.G. (1974). Novyye dannyye of flore Zemli Frantsa-Iosifa. [New data on the flora of Zemlya Frantsa-Iosifa.] *Bot. Zhurn.* **59** (2), 275–9.
Uranov, A.A. (1965). Fitogennoye pole. [The phytogenic field.] *Problemy sovremennoy botaniki* [Problems of contemporary botany], vol. 1, 251–4. Moscow, Leningrad.
Warming, E. (1912). Structure and biology of arctic flowering plants. Saxifragaceae. *Medd. Grønland* **36**, 169–236.
Young, S.B. (1971). The vascular flora of St. Lawrence Island with special reference to floristic zonation in the Arctic regions. *Contrib. Gray Herb. Harvard University*, no. 201, 11–115.
Yurtsev, B.A. (1966). Gipoarkticheskiy botaniko-geograficheskiy poyas i proiskhozhdeniye yeye flory. [The hyparctic botanical-geographical belt and its productivity.] 93 pp. Moscow, Leningrad.
Yurtsev, B.A. (1968). *Flora Suntar-Khayata.* [The flora of Suntar-Khayata.] 235 pp. Leningrad.
Yurtsev, B.A. (1975). Nekotoryye tendentsii razvitiya metoda konkretnykh flor. [Some tendencies for the development of methods of concrete floras.] *Bot. Zhurm.* **60** (1), 69–83.
Yurtsev, B.A. (1977). O sootnoshenii arkticheskoy i vysokogornykh subarkticheskikh flor. [On the correlation between arctic and high-alpine subarctic floras.] In: *Problemy ekologii, geobotaniki, botanicheskoy geografii i floristiki* [Problems of ecology, geobotany, botanical geography and floristics], pp. 125–38. Leningrad.
Yurtsev, B.A. and Semkin, B.I. (1980). Izucheniye konkretnykh i partsialnykh flor c pomoshch'yu matematicheskikh metodov. [The study of concrete and partial floras by means of mathematical methods.] *Bot. Zhurn.* **65** (12), 1706–18.
Yurtsev, B.A., Tolmachev, A.I. and Rebristaya, O. (1978). Floristicheskoye ogranicheniye i razdeleniye Arktiki. [Floristic delimitation and division into areas of the Arctic.] *Arkticheskaya floristicheskaya oblast'* [The Arctic floristic region], pp. 9–104. Leningrad.
Zhukova, A.L. (1972). K flore pechenochnykh mkhov ostrovo Kheysa, Solisberi i Gukera (Zemlya Frantsa-Iosifa). [To the liverwort flora of Hayes, Salisbury and Hooker Islands (Zemlya Frantsa-Iosifa).] *Nov. syst. plant. non-vasc.*, vol. 9, 307–10. Leningrad.
Zhukova, A.L. (1973*a*). Vidovoy sostav i raspredeleniye pechenochnykh mkhov v rastitel'nykh soobshchestvakh rayona Taymyrskogo statsionara. [The specific

composition and distribution of liverworts in the plant associations of the area around the Taymyr Experimental Station.] *Biogeotsenozy Taymyrskogy tundry i ikh produktivnost'* [Biogeocenoses of the Taymyr tundra and their productivity], no. 2, 120–7. Leningrad.

Zhukova, A.L. (1974*b*). Pechenochnyye mkhi o. Rudol'fa (arkhipelag Zemlya Frantsa-Iosifa). [The liverworts of Rudolf' Island (in the archipelago of Zemlya Frantsa-Iosifa).] *Nov. syst. plant. non-vasc.* vol. 10, 272–7. Leningrad.

Zhukova, A.L. (1973*c*). Pechenochnyye mkhi polyarnykh pustyn' Zemli Frantsa-Iosifa. [The liverworts of the polar deserts of Zemlya Frantsa-Iosifa.] Autoref. Doct. thesis. 21 pp. Leningrad.

Zhukova, A.L. (1973*d*). Floristicheskiy analiz pechenochnykh mkhov, Hepaticae, arkhipelaga Zemlya Frantsa-Iosifa. [Floristic analysis of the Hepaticae of Zemlya Frantsa-Iosifa.] *Bot. Zhurn.* **58** (4), 528–39.

Zhukova, P.G. (1967). Chisla khromosom nekotorykh vidov rasteniy Kraynego-severo vostoka SSSR. [Chromosome numbers of some plant species in the extreme northeast of the USSR.] *Bot. Zhurn.* **52**, 1511–16.

Zubkov, A.I. (1934). Kratkiy predvaritel'nyy otchet o geobotanicheskikh rabotakh na Severnom Ostrove Novoy Zemli. [Short preliminary list of geobotanical papers on the North Island of Novaya Zemlya.] Institute of Agriculture of the Far North, typescript.

Latin plant names

Festuca brachyphylla Schult. 56
F. hyperborea Holmen 56, 126, 151
Gastrolychnis apetala (L.) Tolm. et Kozh. 57
Juncus biglumis L. 56, 200
Lagotis minor (Willd.) Standl. 59
Luzula confusa Lindb. 35, 56, 63, 64, 75, 77, 123, 163, 164, 200
L. nivalis Laest. 56, 200
Minuartia macrocarpa (Pursh) Ostenf. 57, 126, 151
M. rubella (Wahlenb.) Hiern 57, 94, 123, 134, 135, 151, 163, 166
Myosotis asiatica (Vesterg.) Schischk. et Serg. 59, 151
Nardosmia frigida (L). Hook. 59, 153, 201
Novosieversia glacialis (Adam) F. Bolle 59, 126, 129, 138
Oxyria digyna (L). Hill 57, 124
Papaver polare (Tolm.) Perf. 34, 37, 57, 64, 70, 73, 75, 76, 78, 82, 83, 86, 93, 94, 98, 99, 100, 103, 108, 123, 132–6, 140, 147, 152, 163, 167, 168, 170, 192–5, 206
P. radicatum Rotbb. 57
Parrya nudicaulis (L.) Regel 57, 126, 151
Phippsia algida (Soland.) R. Br. 15, 34, 37, 41, 49, 51, 56, 63, 64, 70, 73–5, 78, 80, 82–6, 92–4, 97–103, 106–20, 122, 123, 132–4, 136, 140, 141, 144, 147, 148, 159–61, 163, 192, 194, 206
P. concinna (Th. Fries) Lindeb. 56
Pleuropogon sabinii R. Br. 55, 151
Poa abbreviata R. Br. 55, 64, 74, 75, 93, 94, 97, 123, 136, 157, 158, 161, 163, 171, 192, 193
P. alpigena (Fr.) Lindm. 55, 63, 64, 123, 124, 136, 140, 156, 163
P. alpigena (Fr.) Lindm. var. *colpodea* (Th. Fries) Scholand. 55
P. arctica R. Br. 55, 152
P. arctica R. Br. var. *vivipara* Hook. 55
P. lindebergii Tzvel. 55, 151
P. tolmatchewii Roshev. 55
Polygonum viviparum L. 57, 151
Potentilla hyparctica Malte 36, 59, 152
P. pulchella R. Br. 59, 151
Puccinellia angustata (R. Br.) Rand et Redf. 56, 133, 140, 151
P. phryganodes (Trin.) Scribn. et Merr. 56, 151
P. vahliana (Liebm.) Scribn. et Merr. 51, 56
Ranunculus hyperboreus Rottb. 57
R. nivalis L. 57, 153
R. sabinii R. Br. 57, 153, 154, 155
R. sulphureus Soland. 57, 136, 152, 153
Sagina intermedia Fenzl 57
Salix arctica Pall. 57

S. lanata L. 12
S. polaris Wahlenb. 13, 35, 36, 56, 121, 135, 136, 137, 138, 141, 150, 152, 206
S. reptans Rupr. 57, 135, 136
Saussurea tilesii (Ldb.) Ldb. 59, 126, 151, 153
Saxifraga caespitosa L. 59, 64, 94, 123, 132, 133, 134, 135, 136, 163, 189, 190, 192, 193, 200
S. caespitosa L. ssp. *exaratoides* (Simm.) Engl. et Irmsch. 51, 53, 59, 152, 155
S. cernua L. 41, 59, 64, 70, 75, 78, 84, 86, 94, 98, 100, 103, 108, 111, 115, 116, 117, 118, 120, 123, 132, 133, 134, 135, 136, 140, 141, 147, 152, 163, 185, 186, 187, 188, 192, 193, 194, 206
S. foliolosa R. Br. 58, 144, 152
S. hirculus L. 58
S. hyperborea R. Br. 34, 37, 59, 64, 70, 73, 75, 78, 82–4, 86, 94, 98, 100, 103, 108, 111, 115, 117, 118, 123, 163, 188, 189, 193, 194, 206
S. nivalis L. 58, 75, 94, 123, 134, 135, 147, 163, 182, 183, 185, 192
S. oppositifolia L. 41, 59, 64, 70, 100, 123, 124, 132–4, 136, 138, 140, 147, 163, 191, 192, 198
S. platysepala (Trautv.) Tolm. 59, 155
S. pulvinata 191
S. rivularis L. 59, 151
S. serpyllifolia Pursh 59, 126, 128
S. tenuis (Wahlenb.) H. Smith 58, 184, 192
Senecio atropurpureus (Ldb.) B. Fedtsch. 59, 126, 151
Silene acaulis (L.) Jacq. 51, 53, 54, 57
Stellaria ciliatosepala Trautv. 57, 151, 200
S. crassipes Hult. 57
S. edwardsii R. Br. 57, 64, 75, 82, 94, 98, 100, 103, 108, 111, 123, 132, 133, 135, 136, 140, 143, 147, 148, 163, 167
S. humifusa Rottb. 57, 151
S. laeta Richards, 57, 151

Mosses
Andreaea rupestris Hedw. 43, 73, 86, 90, 92, 108, 109, 147
A. rupestris Hedw. var. *acuminata* (B.S.G.) Sharp 44
Aulacomnium turgidum (Wahlenb.) Schwaegr. 46, 47, 64, 73, 78, 98, 143, 144, 145, 148
Bartramia ithyphylla Brid. 45, 65, 78, 86, 94, 100
Bryoerythrophyllum ferruginascens (Stirt.) Giac. 65, 86
Bryum Hedw. 47, 76, 93, 109, 119, 132, 140
B. arcticum (R. Br.) B.S.G. 46, 75
B. nitidulum Lindb. 202

Gymnomitrium concinnatum Corda 78, 87, 108
G. coralloides Nees 87
G. obtusum (Lindb.) Pears. 87
Jungermannia polaris Lindb.
Lophozia alpestris (Schleich.) Evans 46, 65, 75, 78, 104, 11, 116
L. excisa (Dickls.) Dum. 66, 75, 78, 87, 95, 100, 104, 111, 115
L. grandiretis (Lindb.) Schiffn. 78, 104
L. longidens (Lindb.) Macoun var. *arctica* (Schust.) Schljak. 95
L. major (C. Jens.) Schljak. 66, 75
Othocaulis atlanticus (Kaal.) Buch 66
O. kunzeanus (Hüb.) Buch 100
O. quadrilobus (Lind.) Buch 100
Scapania calcicola (Anr. et Pears.) Ingham 66, 75, 78, 87, 95, 104, 111, 115
S. globulifera (C. Jens.) Schljak. 78, 95, 104
S. gymnostomophila Kaal. 75
S. lingulata Buch 75
S. mucronata Buch 95
S. praetervisa Meyl. 95
Sphenolobus minutus (Crantz) Steph. 66, 75, 87, 92, 95, 104
Tritomaria heterophylla Schust. 66
T. quinquedentata (Huds.) Buch 66
T. scitula (Tayl.) Jørg. 66, 78, 87, 95, 100, 111, 115

Lichens

Alectoria jubata (L.) Ach. var. *chalybeiformis* Roumeg. 66, 87
A. minuscula (Nyl. ex Arn.) Degel. 44, 87, 92, 201
A. nigricans (Ach.) Nyl. 43, 44, 66, 73, 74, 75, 87, 92, 95, 100, 104, 201
A. nitidula Th. Fr. 66
A. ochroleuca (Hoffm.) Mass. 43, 44, 66, 75, 87, 95, 100, 108
A. pubescens (L.) Howe 43, 44, 66, 73, 87, 90, 91, 92, 95, 201
Arctomia delicatula Th. Fr. 202
A. interfixa (Nyl.) Vain. 202
Bilimbia subfuscula (Nyl.) Th. Fr. 202
Buellia D. Not. 76, 88, 96, 111
B. coniops (Wahlenb. ex Ach.) Th. Fr. 43, 45, 201
B. punctata (Hoffm.) Mass. 43, 45, 201
Caloplaca elegans (Link.) Th. Fr. 88
C. jungermanniae (Vahl.) Th. Fr. 43, 45, 67, 96, 98, 101
C. scillistidiorum Lynge 67, 96, 101
C. subolivacea (Th. Fr.) Lynge 79, 101, 111
Candelariella vitellina (Ehrh.) Müll. 76, 89, 101, 105, 108
Cetraria cucullata (Bell.) Ach. 34, 37, 42, 45, 47, 48, 63, 66, 69, 70–4, 76, 77, 78,

87, 92, 95, 98, 100, 104, 108, 122, 132, 136, 202, 206
C. delisei Th. Fr. 32, 33, 66, 79, 82, 83, 87, 98, 99, 101, 102, 104, 111, 117, 122, 125, 132–4, 136, 140, 144, 147, 202
C. elenkinii Krog 202
C. ericetorum Opiz 34, 45, 47, 48, 66, 69, 70, 73, 74, 76, 77, 87, 92, 95, 98, 101, 104
C. hepatizon (Ach.) Vain. 44, 66, 87
C. islandica (L.) Ach. 136
C. islandica (L.) Ach. var. *polaris* Rassad. 32, 66, 79, 87, 92, 105, 147, 202
C. nigricans (Retz.) Nyl. 87
C. nivalis (L.) Ach. 45, 47, 48, 66, 69, 76, 77, 79, 87, 95, 101, 105, 108
C. simmonsii Krog
Cladonia elongata (Jack.) Hoffm. 201
C. pocillum (Ach.) Rich. 67, 79, 88, 101, 105, 111, 149
C. pyxidata (L.) Hoffm. 67, 76, 79, 85, 88, 92, 98, 101, 105, 108, 111, .112, 115
Collema ceraniscum Nyl. 34, 45, 46, 47, 48, 67, 73, 79, 80, 82, 83, 96, 98, 101, 105, 108, 111, 202
Cornicularia aculeata (Schreb.) Ach. 87
C. divergens Ach. 43, 44, 66, 74, 76, 87, 92, 101, 105
Dactylina Nyl. 152
D. arctica (Hook.) Nyl. 202
D. madreporiformis (Ach.) Tuck.
Gasparrinia arctica Golubk. et Savicz 121
Haematomma ventosum (L.) Mass. 201
Hypogymnia intestiniformis (Vill.) Räs. 44, 98
H. oroarctica Krog. 202
H. physodes (L.) Hyl. 88
Lecanora campestris (Schaer.) Hue 43, 45, 76, 79, 88, 96, 97, 108
L. epibryon Ach. 67, 201
L. polytropa (Ehrh.) Rabenh. 43, 44, 88, 201
Lecidea (Ach.) Mass. 67, 96, 98, 101, 105, 108, 201
L. cyanea (Ach.) Röhl. 37
L. dicksonii (Gmel.) Ach. 38, 43, 44, 73, 88, 92, 108, 109
L. macrocarpa (DC.) Steud. 43, 44, 73, 88, 92
Leciophysma finnmarkicum Th. Fr. 202
Lepraria Ach. 67
L. arctica (Lynge) Wetmore 202
Lopadium coralloideum (Nyl.) Lygne 202
Neuropogon sulphureus (Koenig) Elenk. 130, 201
Ochrolechia Massal. 15, 34, 42, 80, 91, 96, 105, 108, 129
O. frigida (Sw.) Lynge 33, 43, 45, 47, 67,

224   List of Latin plant names

73, 76, 79, 89, 92, 96, 98, 101, 111, 115,
119, 130, 145, 201, 202, 206
O. gonatodes (Ach.) Räs. 203
O. grimmiae Lynge 203
O. tartarea (L.) Mass. 41, 67, 79, 83, 89,
101, 105, 111, 120
Pannaria pezizoides (Web.) Trevis 43, 45,
67, 73, 89, 144, 202
Parmelia omphalodes (L.) Ach. 67, 79, 88,
96, 202
Parmeliella arctophila (Th. Fr.) Malme 202
Peltigera erumpens (Th. Tayl.) Vain. 105
P. rufescens (Weis) Humb. var. *incusa*
(Flot.) Koerb. 67, 88
Pertusaria DC. 15, 34, 42, 50, 67, 73, 76,
79, 80, 82, 83, 89, 91, 92, 96, 98, 101,
105, 108, 111, 129, 145
P. bryontha (Ach.) Nyl. 50
P. bryophaga Erichs.
P. dactylina (Ach.) Nyl. 50
P. freyii Erichs. 47, 79
P. glomerata (Ach.) Schaer. 33, 45, 47, 50,
79, 130, 202, 206
P. octomela (Norm.) Erichs. 79
P. oculata (Dicks.) The. Fr. 50
P. subdactylina Nyl. 50, 202
Physcia muscigena (Ach. Nyl. 67
Placodium subfruticulosum Elenkin 202
Psoroma hypnorum (Vahl.) S. Gray 43, 45,
67, 143, 144, 201, 202, 203
Rhizocarpon geographicum (L.) DC. 42, 43,
44, 88, 92, 201
Rinodina turfacea (Ach.) Koerb. 43, 44, 67,
73, 79, 89, 92, 98, 101, 201, 203
Siphula ceratites (Wahlenb.) Fr. 202
Solorina bispora Nyl. 67, 98, 101, 105
Sphaerophorus fragilis (L.) Pers. 87

S. globosus (Huds.) Vain. 44, 66, 73, 76,
87, 140, 201, 202
Stereocaulon alpinum Laur. 66, 76, 79, 88,
92, 95, 105
S. botryosum Ach. em. Frey f. *congestum*
(H. Magn.) Frey 87
S. rivulorum H. Magn. 41, 42, 46, 48, 49,
66, 76, 79, 80, 82–5, 88, 95, 98, 99, 101,
102, 105–11, 113–18, 120, 121–3, 144,
147–9, 202, 206
S. vesuvianum Pers. var. *depressum* (H.
Magn.) M. Lamb 42, 43, 45, 93, 95, 97
S. vesuvianum Pers. var. *pseudofastigiatum*
M. Lamb 66, 73, 88
Thamnolia Schaer. 74, 130, 138, 145, 152
T. subuliformis (Ehrh.) W. Culb. 41, 130,
144, 145, 147, 148, 149, 202
T. vermicularis (Sw.) Ach. ex Schaer. 67,
70, 72, 73, 76, 79, 82, 88, 96, 98, 101,
105, 108, 130, 133, 145
Toninia candida (Web.) Th. Fr. 201
T. lobulata (Sommerf.) Vain. 50, 128, 129,
147, 149, 202
Umbilicaria Hoffm. em. Frey 42, 43, 90,
91, 97, 122
U. arctica (Ach.) Nyl. 43, 44, 88
U. cylindrica (L.) Del. 43
U. decussata (Vill.) Frey 43
U. hyperborea (Ach.) 43
U. proboscidea (L.) Schard. 43, 44, 67, 73,
88, 90, 92, 96, 101, 206
Usnea sulphurea (Koenig) Th. Fr. 43, 44,
50, 73, 88, 90, 92, 101, 102, 108, 109,
122, 130, 201
Verrucaria aethiobola Wahlenb. 76, 89, 96

Blue-green algae
Nostoc commune Vauch. 80, 120, 136

Index

Geographical names and Latin names of plant families are fully indexed while concepts and terms are referred only to the pages where they are defined or discussed in depth. Page numbers in bold refer to maps, those in italic to tables.